Air and Water Pollution XXX

WITPRESS

WIT Press publishes leading books in Science and Technology.
Visit our website for the current list of titles.
www.witpress.com

WITeLibrary

Home of the Transactions of the Wessex Institute.
Papers contained in this volume are archived in the WIT eLibrary in volume 259 of WIT
Transactions on Ecology and the Environment (ISSN 1743-3541).
The WIT eLibrary provides the international scientific community with immediate and
permanent access to individual papers presented at WIT conferences.
Visit the WIT eLibrary at www.witpress.com.

THIRTIETH INTERNATIONAL CONFERENCE ON
MODELLING, MONITORING AND MANAGEMENT OF AIR
AND WATER POLLUTION

Air and Water Pollution 2022

CONFERENCE CHAIRMEN

Stefano Mambretti
Polytechnic of Milan, Italy
Member of WIT Board of Directors

James Longhurst
University of the West of England, UK

Joanna Barnes
University of the West of England, UK

INTERNATIONAL SCIENTIFIC ADVISORY COMMITTEE

Angela Baeza Serrano
Daniel Bonotto
Zuzana Boukalova
Miguel De Luque
Israel Felzenszwalb
Joana Ferreira
Eric Gielen
Jabulani Gumbo
Rosaria Ippolito
Piotr Kowalski
Myriam Lopes
Julia Lu
Elena Magaril
Robert Mahler
Florica Manea
Guido Marseglia

Ana Isabel Miranda
Shiva Nagendra
Deborah Panepinto
Rene Parra
Aniela Pop
Elena Rada
Marco Ravina
Angelo Riccio
Irina Rukavishnikova
Marco Schiavon
Alexander Slobodov
Jose Antonio Souto Gonzalez
Erick Giovani Sperandio Nascimento
Carlo Trozzi
Ben Williams
Giuseppe Zappala

Organised by
Wessex Institute, UK
University of The West of England, UK
University of Santiago de Compostela, Spain

Sponsored by
WIT Transactions on Ecology and the Environment
International Journal of Environmental Impacts

WIT Transactions

Wessex Institute
Ashurst Lodge, Ashurst
Southampton SO40 7AA, UK

We would like to express thanks to all the conference Chairs and members of the International Scientific Advisory Committees for their efforts during the 2022 conference season.

Juraj Muzik University of Zilina, Slovakia

Richard Mwaipungu Sansutwa Simtali Ltd, Tanzania

Shiva Nagendra Indian Institute of Technology Madras, India

Fermin Navarrina University of A Coruña, Spain

Norwina Mohd Nawawi International Islamic University Malaysia, Malaysia

Derek Northwood University of Windsor, Canada

David Novelo-Casanova National Autonomous University of Mexico, Mexico

Andrzej Nowak Silesian University of Technology, Poland

Freeman Ntuli Botswana International University of Science and Technology, Botswana

Miguel Juan Nunez-Sanchez European Maritime Safety Agency, Portugal

Suk Mun Oh Korea Railroad Research Institute, South Korea

Yasuo Ohe Tokyo University of Agriculture, Japan

Roger Olsen CDM Smith, USA

Antonio Romero Ordonez University of Seville, Spain

Francisco Ortega Riejos Universidad de Sevilla, Spain

Ozlem Ozcevik Istanbul Technical University, Turkey

Leandro Palermo Jr University of Campinas, Brazil

Deborah Panepinto Turin Polytechnic, Italy

Marilena Papageorgiou Aristotle University of Thessaloniki, Greece

Jose Paris University of A Coruna, Spain

Bum Hwan Park Korea National University of Transportation, South Korea

Bekir Parlak Bursa Uludag University, Turkey

Rene Parra Universidad San Francisco de Quito, Ecuador

Marko Peric University of Rijeka, Croatia

Roberto Perruzza CERN, Switzerland

Cristiana Piccioni Sapienza University of Rome, Italy

Max Platzer AeroHydro Research & Technology Associates, USA

Lorenz Poggendorf Toyo University, Japan

Dragan Poljak University of Split, Croatia

Antonella Pontrandolfi Council for Agricultural Research & Economics, Italy

Aniela Pop Polytechnic University of Timisoara, Romania

Serguei Potapov French Electricity (EDF), France

Maria Pregnolato University of Bristol, UK

Dimitris Prokopiou University of Piraeus, Greece

Yuri Pykh Russian Academy of Sciences, Russia

Sue Raftery OnPoint Learning, USA

Marco Ragazzi University of Trento, Italy

Marco Ravina Turin Polytechnic, Italy

Jure Ravnik University of Maribor, Slovenia

Joseph Rencis California State Polytechnic University, USA

Genserik Reniers University of Antwerp, Belgium

Admilson Irio Ribeiro São Paulo State University, Brazil

Jorge Ribeiro University of Lisbon, Portugal

Angelo Riccio University of Naples "Parthenope", Italy

Corrado Rindone Mediterranea University of Reggio Calabria, Italy

German Rodriguez Universidad de Las Palmas de Gran Canaria, Spain

Rosa Rojas-Caldelas Autonomous University of Baja California, Mexico

Jafar Rouhi University of Campania "L. Vanvitelli", Italy

Irina Rukavishnikova Ural Federal University, Russia

Francesco Russo University of Reggio Calabria, Italy

Shahrul Said Universiti Teknologi MARA, Malaysia

Hidetoshi Sakamoto Doshisha University, Japan

Seddik Sakji INFRANEO, France

Artem Salamatov Chelyabinsk State University, Russia

Daniel Santos-Reyes ICHI Research & Engineering, Mexico

Bozidar Sarler University of Ljubljana, Slovenia

Martin Schanz Graz University of Technology, Austria

Evelia Schettini University of Bari, Italy

Air and Water Pollution XXX

Editors

Stefano Mambretti
Polytechnic of Milan, Italy
Member of WIT Board of Directors

James Longhurst
University of the West of England, UK

Joanna Barnes
University of the West of England, UK

WITPRESS Southampton, Boston

Editors:

Stefano Mambretti
Polytechnic of Milan, Italy
Member of WIT Board of Directors

James Longhurst
University of the West of England, UK

Joanna Barnes
University of the West of England, UK

Published by

WIT Press
Ashurst Lodge, Ashurst, Southampton, SO40 7AA, UK
Tel: 44 (0) 238 029 3223; Fax: 44 (0) 238 029 2853
E-Mail: witpress@witpress.com
http://www.witpress.com

For USA, Canada and Mexico

Computational Mechanics International Inc
25 Bridge Street, Billerica, MA 01821, USA
Tel: 978 667 5841; Fax: 978 667 7582
E-Mail: infousa@witpress.com
http://www.witpress.com

British Library Cataloguing-in-Publication Data

A Catalogue record for this book is available
from the British Library

ISBN: 978-1-78466-467-1
eISBN: 978-1-78466-468-8
ISSN: 1746-448X (print)
ISSN: 1743-3541 (on-line)

The texts of the papers in this volume were set individually by the authors or under their supervision. Only minor corrections to the text may have been carried out by the publisher.

Preface

This conference is the merging of two successful events organised by the Wessex Institute, the International Conference on Modelling, Monitoring and Management of Air Pollution, which started in Mexico in 1993; and the International Conference on Monitoring, Modelling and Management of Water Pollution, which originated in Southampton, UK in 1991.

This volume contains papers presented at the Conference which, scheduled in Milan, Italy, was held on line due to the Coronavirus pandemic. The Conference has been organised by the Wessex Institute, in collaboration with the Politecnico di Milano in Italy and University of the West of England in the UK.

The conference brought together contributions from scientists from around the world to present recent work on various aspects of the air pollution phenomena. Discussion ranged across the terrain of air quality modelling, monitoring and management with specific sessions covering modelling studies, indoor air pollution, emission studies, air pollution management, policy and legislation and monitoring approaches. Moreover, engineers and scientists working in water pollution must be familiar with a wide range of issues including the physical processes of mixing and dilution, chemical and biological processes, mathematical modelling, data acquisition and measurement, to name but a few. In view of the scarcity of available data, it is important that experiences are shared on an international basis. Thus, a continuous exchange of information between scientists from different countries is essential.

The availability of unlimited resources cannot any longer be taken for granted as the needs of a growing world population, demanding better standards of living, continues to increase. The scientific knowledge derived from well-designed studies needs to be allied with further technical and economic studies in order to ensure cost-effective and efficient mitigation. In turn, the science, technology and economic outcomes are necessary but not sufficient. Increasingly, the conference has recognised that the outcome of such research needs to be contextualised within well-formulated communication strategies that help policymakers and citizens to understand and appreciate the risks and rewards arising from air and water pollution management. Consequently, the Conference has enjoyed a wide range of high-quality papers that develop the fundamental science of pollution and an equally impressive range of presentations that places these new developments within the frame of its mitigation and management.

Environmental problems are essentially interdisciplinary and both series, which now are joint in order to assure a more comprehensive view of the matter, have provided a forum for discussion amongst scientists, managers and academics from different areas of contamination. There remains much to do and the conference series will continue to play an important role in providing an opportunity for an international audience to discuss both long-standing and emergent issues in air pollution science and policy.

The papers selected for presentation and published in this volume are part of the Transactions on Ecology and the Environment. They have been archived online in the WIT eLibrary (www.witpress.com/elibrary) where they are easily and permanently available to the international scientific community. These collected papers provide an invaluable record of the development of science and policy pertaining to air pollution.

The Editors wish to thank the authors for their contributions to the conference and to acknowledge the assistance of the eminent members of the International Scientific Advisory Committee for their support for the conference and in particular for their peer reviewing of the manuscripts.

The Editors

Contents

Section 1: Atmospheric modelling and forecasting

Assessment of cumulus parameterization schemes in modeling
meteorology associated with an air pollution event in the Andean
region of Ecuador
Rene Parra.. 3

Nutrients in marginal land soils and their potential effect on the
environment
Nicole Rodriguez, Timothy Ho, Zhongwei Shi & Julia Lu................................. 15

Ensemble deep learning for classification of pollution peaks
Phuong N. Chau, Rasa Zalakeviciute & Yves Rybarczyk................................. 25

Use of satellite images to estimate urban heat maps
*Ana Gabriela Fernández-Garza, Eric Gielen
& José-Sergio Palencia-Jiménez*... 37

Section 2: Global, regional and local studies

Importance of comprehensive health risk assessment procedures for
modern waste-to-energy facilities in complex geographical contexts
oriented to circular economy
Elena Cristina Rada, Marco Tubino, Marco Schiavon & Luca Adami............................. 53

Environmental impact evaluation of odor dispersion emitted from pig
farms: A case study from Alban, Cundinamarca, Colombia
*Gerardo Romero-Tovar, Hernan D. Granda-Rodriguez
& Miguel A. de Luque-Villa*.. 65

Local monitoring of traffic-related air pollution around schools in
south east London, UK
Ho Yin Wickson Cheung & Liora Malki-Epshtein... 75

Evaluation of the ecological state using the water quality index and
fluvial habitat index of the urban basins of Panama
*Quiriatjaryn M. Ortega-Samaniego, Andres Fraiz, Arturo Dominici,
Haydee Osorio, Adrian Ramos-Merchante, Edgar Arauz, Maria Paches
& Inmaculada Romero* 87

Assessment of surface and groundwater contamination and seasonal
variation at the tannery area in Dhaka, Bangladesh
*Asia Akter, Afrose Sultana Chamon, Md. Nadiruzzaman Mondol
& Syed Mohammed Abul Faiz* 99

Water management in Colombia from the socio-ecological systems
framework
Miguel A. de Luque-Villa & Mauricio González-Méndez 111

Section 3: Water treatment

Design of the interceptor-collector and wastewater treatment system
for pollution mitigation: A case study
*Bethy Merchán-Sanmartin, Paúl Carrión-Mero, Fernando Morante-Carballo,
Josué Briones-Bitar, Adrián Gonzalez-Rugel & Hairo Vera-Demera* 125

Evolution of the activated sludge community of a wastewater treatment
plant with industrial discharges
*Ángela Baeza-Serrano, Feliu Sempere, Nuria Oliver, Pilar Gutiérrez
& Gloria Fayos* 137

Holistic approach to the economic benefits of using reclaimed water
in agriculture
María José López Serrano 147

Author index 157

SECTION 1
ATMOSPHERIC MODELLING
AND FORECASTING

ASSESSMENT OF CUMULUS PARAMETERIZATION SCHEMES IN MODELING METEOROLOGY ASSOCIATED WITH AN AIR POLLUTION EVENT IN THE ANDEAN REGION OF ECUADOR

RENE PARRA
Instituto de Simulación Computacional, Colegio de Ciencias e Ingeniería,
Universidad San Francisco de Quito, Ecuador

ABSTRACT

Air quality results from the interaction between emissions and meteorology. When significant changes occur, such as a sudden drop in temperature, atmospheric stability can persist, promoting air pollution. Between 7 and 9 November 2020, there was a significant drop in temperature in Cuenca (2,500 masl), a city located in the Andean zone in southern Ecuador. At noon, decreases of up to 700 W m^{-2} and 10°C were recorded in the solar radiation and temperature levels, respectively, compared to the first days of the month. At the same time, increments of around 20 μg m^{-3} in $PM_{2.5}$ hourly concentrations were recorded. Cumulus convection is a process directly related to cloud formation and, therefore, to solar radiation and temperature. We simulated the meteorology from 2 to 11 November 2020, with the Weather Research and Forecasting model (WRF 4.0.3), with a resolution of 1 km, without cumulus parameterization, and with nine options for this component. Modeling without this parameterization provided acceptable results for solar radiation and temperature. However, this option overestimated wind speed at the surface. Globally, option 10 (Kain–Fritsch Cumulus Potential) presented the best modeling performance. Options 3 (Grell–Freitas) and 6 (Tiedtke) were better or similar than modeling without cumulus parameterization. None of the options adequately modeled the temperature and solar radiation on 9 November 2020, the day on which the lowest values of these variables were observed, suggesting that the model will not provide proper values for days with sudden decreases in solar radiation and temperature. Because most parameterized effects, such as the boundary layer, convection, microphysics, and surface schemes, are closely linked, it is necessary to study their influence. It is also necessary to assess the potential benefit of data assimilation and even the development of dedicated schemes for numerical simulation in the equatorial zone of the Andes.
Keywords: Cuenca, cumulus convective, atmospheric forecasting, air quality forecasting, atmospheric stability.

1 INTRODUCTION

Air quality results from atmospheric emissions and meteorology [1]. When significant changes occur in atmospheric conditions, such as a sudden drop in temperature, atmospheric stability persists, and air pollution increases. The simulation of these events tests the performance of numerical models, which are expected to capture both meteorology and air quality. Air quality forecasting is crucial for reducing exposure and restricting anthropogenic sources during predicted periods of high pollution [2]. For this purpose, meteorological variables involved in atmospheric stability, like solar radiation, temperature, and wind speed at the surface, must be appropriately modeled.

Cumulus convection, which is typically parameterized, is a process directly related to cloud formation and, therefore, to solar radiation and temperature drops at the surface.

WIT Transactions on Ecology and the Environment, Vol 259, © 2022 WIT Press
www.witpress.com, ISSN 1743-3541 (on-line)
doi:10.2495/AWP220011

1.1 The air pollution event in Cuenca in November 2020

Between 2 and 11 November 2020, there was a substantial variation in cloudiness in Cuenca (2,500 masl), a city located in the Andean zone of southern Ecuador (Fig. 1). During 1–3 November (Sunday–Tuesday), there was little or no cloudiness, and in the following days, the atmosphere presented a partially or wholly cloudy cover (Fig. 2). On 9 November 2020 (Monday), the cloudiness was persistent. As consequence, decreases of up to 700 W m^{-2} and 10°C were recorded in the solar radiation and temperature levels at noon, respectively, compared to the previous days (Fig. 3). At the same time, the concentrations of primary pollutants increased. At the urban center, increments of around 20 μg m^{-3} in PM$_{2.5}$ hourly concentrations were recorded (Fig. 3).

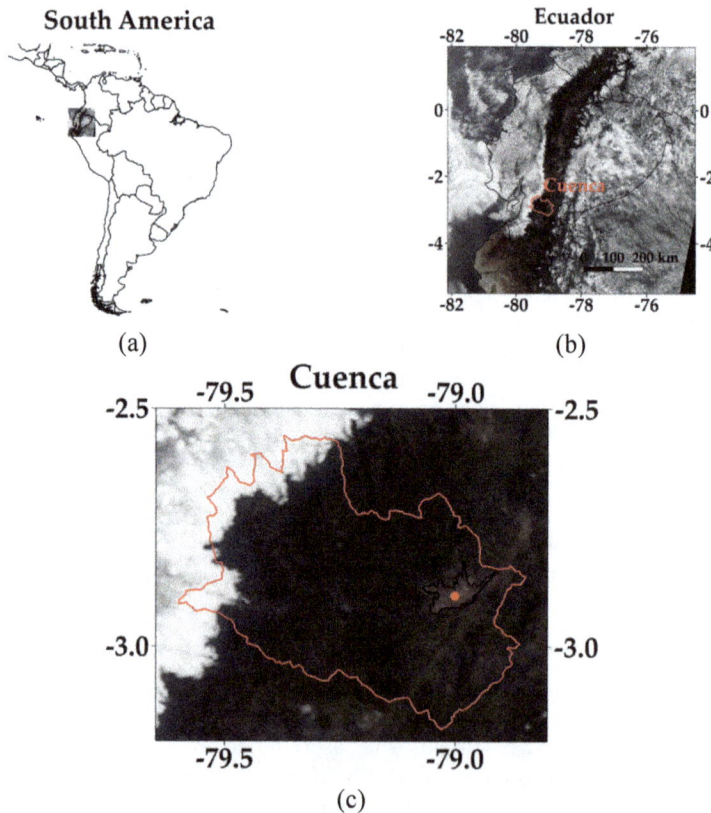

Figure 1:   Location of: (a) Ecuador; and (b) Cuenca. (c) The urban area of Cuenca (black border) and MUN station (red dot, 2,500 masl). Satellite image corresponds to 3 November 2020 (Terra satellite, 10:30 LT) [5].

In the urban area of Cuenca, PM$_{2.5}$ emissions come mainly from vehicles that use diesel [3], [4]. Hourly emissions are typically higher during working days, and there are decreases on Saturdays, especially on Sundays, due to a lower flow of this type of vehicle. We did not identify the influence of other relevant sources during the study period. Fig. 3 indicates

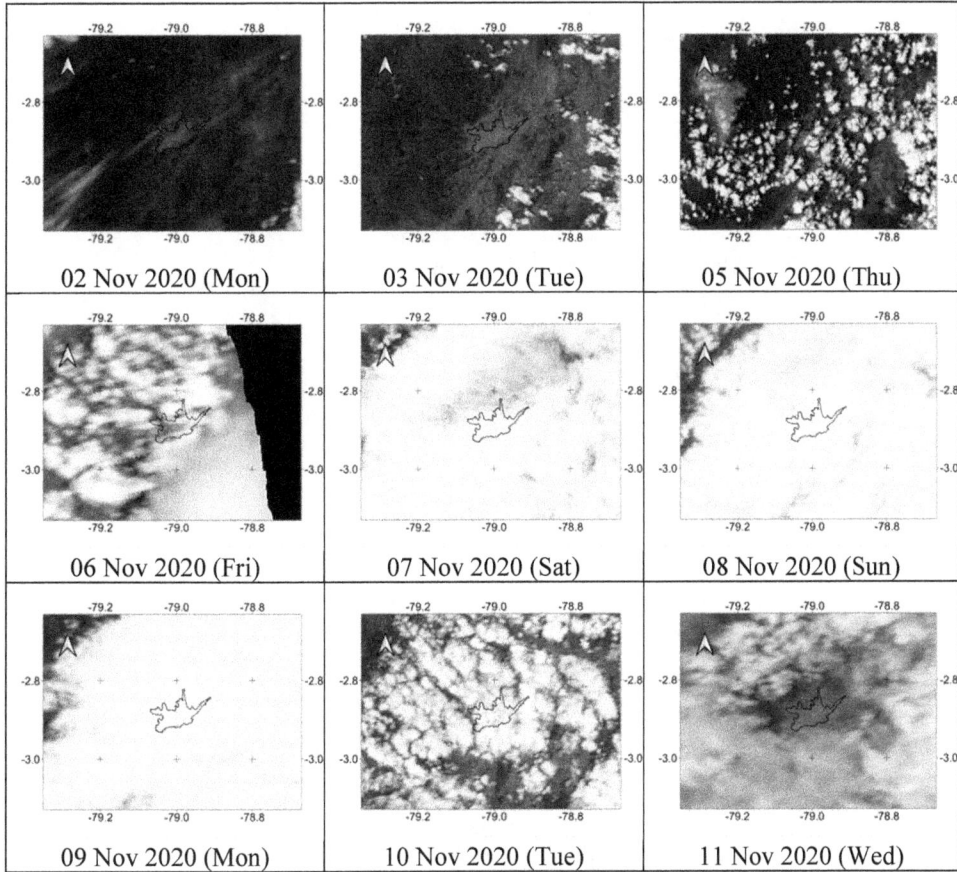

Figure 2:   Satellite images from 2 to 11 November 2020 (Aqua satellite, 13:30 LT) [5]. The black border indicates the urban area of Cuenca.

maximum hourly concentrations of around 40 μg m^{-3} from 2 to 6 November (Monday to Friday) 2020. A drop in temperature is observed from 7 November (Saturday). On 9 November (Monday), the maximum temperature (15°C) was about 10°C lower than the previous weekdays. The maximum hourly concentration of PM$_{2.5}$ on Monday 9 November (60 μg m^{-3}) was 58% higher than the maximum hourly concentration on Monday 2 November (38 μg m^{-3}). If the PM$_{2.5}$ emissions were of the same magnitude on these two days, the difference in concentrations is explained by the more significant restriction of the atmosphere on 9 November to disperse the pollutants.

1.2 Modeling in the "grey zone"

Complex topography and land-use heterogeneity directly affect the atmospheric dynamic [6]. At 1 km of resolution, the topography representation of complex terrain improves. This benefit is significant for regions such as Cuenca. For this reason, atmospheric modeling was done using this spatial resolution for studying the influence of planetary boundary layer

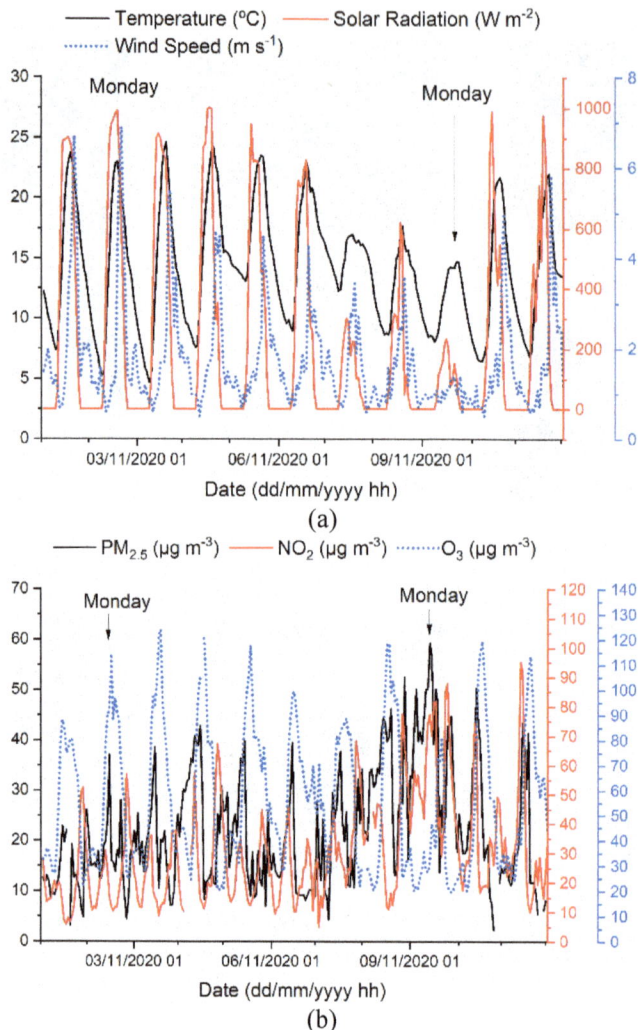

Figure 3: Records of the MUN station during 1 to 11 November 2020. (a) Meteorology; and (b) Air quality.

schemes in Cuenca [7], for assessing the influence on air quality due to the shift from diesel to electric buses [6], and the influence of feedback between aerosols and meteorology [8].

This horizontal grid spacing corresponds to the "grey zone", a range of resolution in which phenomena such as turbulence, convective transport, and clouds are partly resolved, therefore, implying that might not be required modeling with cumulus convective parameterization [9]. Modeling in the grey zone is interesting to address whether working with higher resolution is always better [10].

As the length scale for convection (1 to 10 km) is similar to a spatial resolution of 1 km [9], this contribution explores the influence of cumulus parameterization schemes in modeling temperature, wind speed and solar radiation at the surface, from 2 to 11 November 2020, to address the following questions:

- What were the cumulus parameterization schemes with the best modeling performances?
- Is there a benefit when modeling without cumulus parameterization?
- Were the meteorological variables adequately modeled for 9 November 2020?

2 METHOD

2.1 Modeling approach

We used Weather Research and Forecasting (WRF version 4.0.3) [11] to model meteorology in Cuenca from 2 to 11 November 2020. Simulations were done through a master domain of 70 × 70 cells (27 km each) and three nested subdomains. The cells of the third domain (100 × 82) have 1 km of resolution and cover the region of Cuenca (Fig. 1(c)). Initial and boundary conditions were generated from the final NCEP FNL Operational Global Analysis data [12]. Table 1 summarizes the schemes and options selected for modeling. The variable "cu_physics" is used in WRF to select the cumulus parameterization options.

Table 1: Schemes and options for modeling meteorology in Cuenca (WRF 4.0.3) [13].

Component	Option	Scheme/model
Microphysics (mp_physics)	2	Purdue Lin
Longwave radiation (ra_lw_physics)	1	RRTM
Shortwave radiation (ra_sw_physics)	2	Goddard
Surface layer (sf_clay_physics)	1	Revised MM5
Land surface (sf_surface_physics)	1	5-layer thermal diffusion
Planetary boundary layer (bl_pbl_physics)	1	Yonsei University
Cumulus parameterization (cu_physics)	0	Without cumulus parameterization
	1	Kain–Fritsch
	2	Betts–Miller–Janjic, Janjic
	3	Grell–Freitas
	5	Grell–3D
	6	Tiedtke
	10	Kain–Fritsch–Cumulus potential
	11	Multi-scale Kain–Fritsch
	14	KIAPS SAS
	16	New Tiedtke

2.2 Metrics for modeling performance

Modeled hourly records of temperature and wind speed at the surface were assessed through the following metrics:

$$GE = \frac{1}{N}\sum_{i=1}^{N}|Pi - Oi|, \ BIAS = Pm - Om, \ RMSE = \sqrt{\frac{1}{N}\sum_{i=1}^{N}(Pi - Oi)^2}$$

$$IOA = 1 - \frac{\sum_{i=1}^{N}(Pi - Oi)^2}{\sum_{i=1}^{N}[|Pi - Pm| + |Oi - Om|]}$$

where GE = gross error; RMSE = root mean square error; IOA = index of agreement; N = number of values; Pm = mean modeled value; Om = mean observed value; Pi = modeled value; Oi = observed value. Table 2 indicates the benchmark values for these indicators.

Table 2: Metrics for meteorological modeling [14].

Parameter	Indicator	Benchmark
	GE	$< 2°C$
Hourly surface temperature	BIAS	$< ±0.5°C$
	IOA	≥ 0.8
	RMSE	< 2 m s^{-1}
Hourly wind speed at 10 m above the surface	BIAS	$< ± 0.5$ m s^{-1}
	IOA	≥ 0.6

In addition, modeled hourly values of global solar radiation at the surface were compared with records. For this variable, linear regressions (y = ax + b) were generated, corresponding x to records and y to computed values. In the case of a perfect fit, a and b variables will be 1 and 0, respectively, and the coefficient of determination (R^2) is equal to 1.

Records were provided by the MUN automatic station, which is located in the city's historic center (MUN station, Fig. 1(c)).

3 RESULTS AND DISCUSSION

Fig. 4 shows the hourly solar radiation records, temperature, wind speed at the surface, and the corresponding modeled values without cumulus parameterization (cu_physics = 0). Although their behaviors were captured by modeling, on 9 November 2020 they were overestimated. At 13:00 LT (9 November 2020), modeled solar radiation, temperature, and wind speed were up to 435 W m^{-2}, 5.5°C, and 1.1 m s^{-1} higher that the corresponding records. Similar patterns were obtained when modeling with cumulus parameterization (cu_physics = 1, 2, 3, 5, 6, 10, 11, 14, 16) (Fig. 5).

For the modeled period (2 to 11 November 2020), none of the cumulus options provided GE values for temperature into the benchmark range (GE < 2°C). The cumulus options with the lowest values for GE (2.2°C) were 0, 10, and 16 (Table 3). Options 0, 1, 3, and 10 provided the best BIAS values (–0.1 or 0.0°C) and into the benchmark range (< ±0.5°C). All the options provided IOA values higher than 0.8 (benchmark range). Options 1, 3, 10, and 16 provided the highest IOA values (0.90 or 0.91). Options 0 (no cumulus parametrization) and 10 showed the best metrics values for modeling the surface temperature.

Fig. 6 shows the observed and modeled wind speed values. The options reached RMSE values into the benchmark range (< 2 m s^{-1}) for wind speed. Options 1, 3, and 10 showed the best RMSE (1.1 m s^{-1}). Options 1, 3, 5, 6, 10, 11, and 14 provided the best BIAS values (–0.1, 0.0, or 0.1 m s^{-1}). All the options provided IOA values higher than 0.6 (benchmark range). Options 6, 10, and 14 obtained the highest IOA values (0.73 or 0.75). Options 6 and 10 showed the best metrics values for modeling the wind speed.

Options 1 and 10 obtained the best values for the linear regression between records and computed global solar radiation values (Table 3), with 0.79 to 0.80, 1.11, and 41.48 to 46.89 for R^2, a, and b, respectively. Options 0, 1, and 10 modeled better global solar radiation. Fig. 7 compares the records with modeled levels for the option without cumulus scheme. Even this option overestimated the solar radiation values on 9 November 2020.

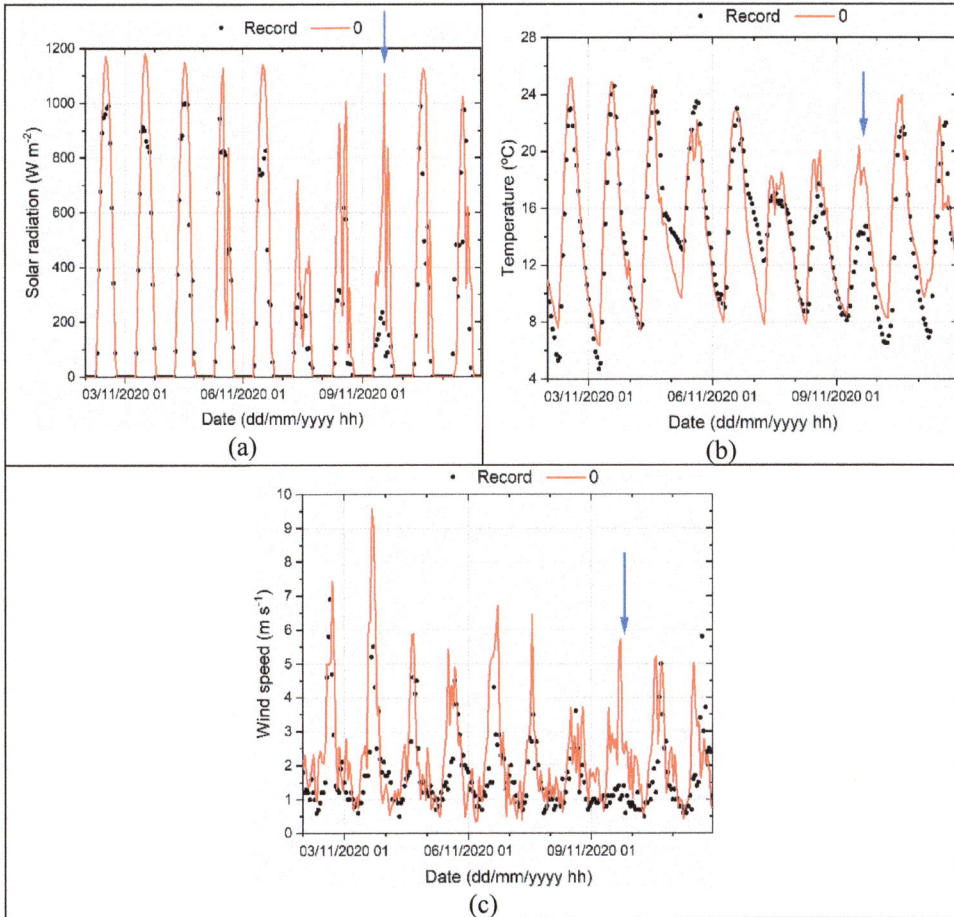

Figure 4: Observed and modeled values without cumulus parameterization (cu_physics = 0). (a) Solar radiation; (b) Temperature at the surface; and (c) Wind speed at 10 m above the surface. Blue arrow indicates midday on 9 November 2020.

Globally, option 10 (Kain–Fritsch Cumulus Potential [15]) presented the best modeling performance. This option replaces the ad hoc trigger function in the Kain–Fritsch cumulus parameterization (cu_physics = 1) with a trigger function related to the distribution of temperature and humidity in the convective boundary layer via probability density functions.

Options 3 (Grell–Freitas [16]) and 6 (Tiedtke [17]) were better or similar than modeling without cumulus parameterization. Option 3 is an improved Grell–Davenye scheme that tries to smooth the transition to cloud-resolving scales [13]. Option 6 is a mass-flux type scheme with convective potential energy removal, time scale, shallow component, and momentum transport.

None of the options adequately modeled the meteorological variables on 9 November 2020, the day on which the lowest values of solar radiation, temperature, and wind speed were observed during the studied period, suggesting that for forecasting purposes, none of

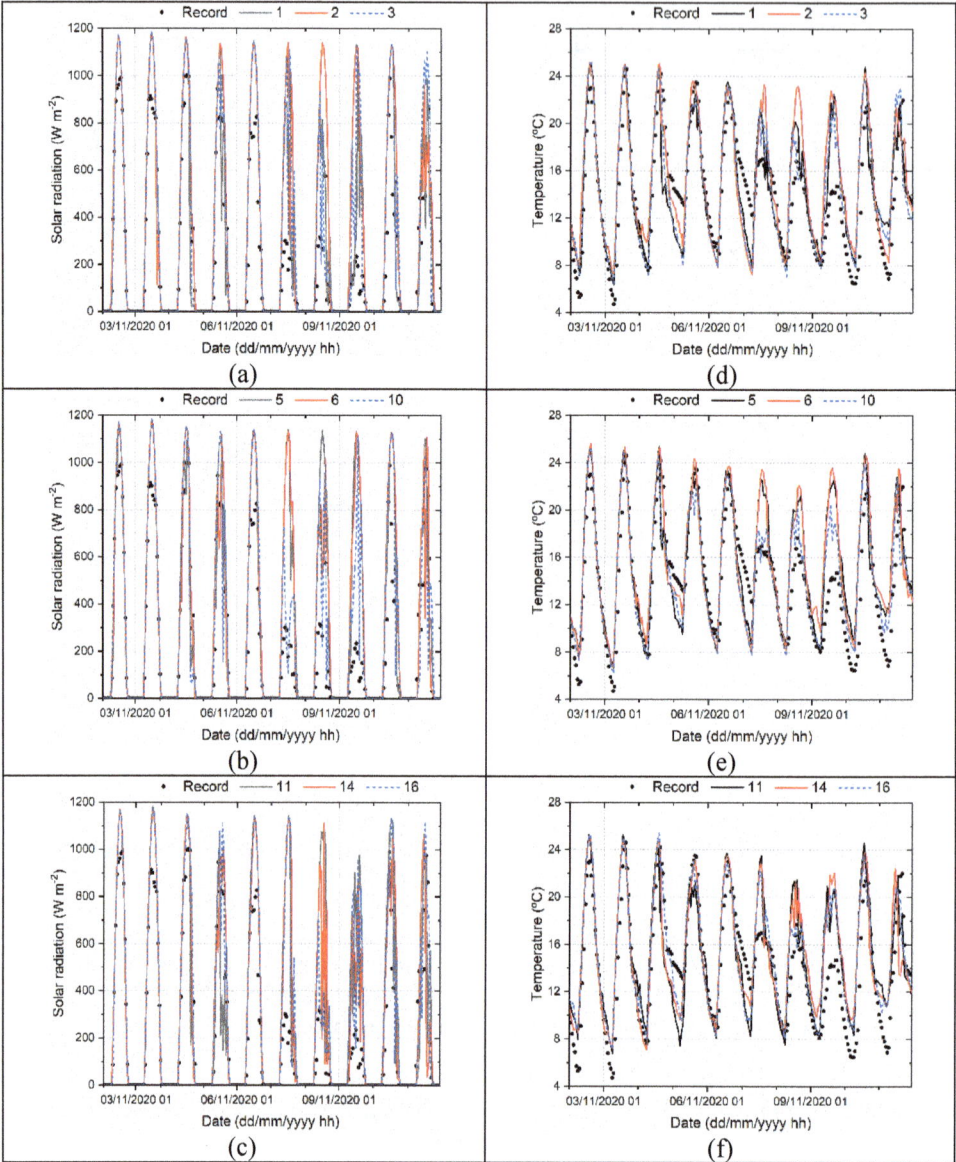

Figure 5:   Observed and modeled values with different cumulus parameterization (cu_physics options: 1, 2, 3, 5, 6, 10, 11, 14, 16). (a), (b), (c) Solar radiation; and (d), (e), (f) Temperature.

the assessed cumulus options will provide proper values for days with significant drops in the magnitude of these parameters.

As 9 November 2020 was an overcast cloudiness day, its atmospheric stability corresponds to the neutral category D of the categories proposed by Pasquill [18]. The

Table 3:   Metrics for meteorological modeling. Bold numbers fit the benchmark criterion. Numbers with grey background indicates the options with the best performances.

	Benchmark [14]	Cu_physics options									
		0	1	2	3	5	6	10	11	14	16
Hourly surface temperature:											
GE	< 2°C	2.2	2.4	2.5	2.4	2.5	2.6	2.2	2.6	2.5	2.2
BIAS	< ±0.5°C	−0.1	0.0	0.6	−0.1	0.6	0.8	−0.1	0.3	0.4	0.4
IOA	≥ 0.8	0.91	0.89	0.88	0.90	0.88	0.88	0.91	0.88	0.88	0.91
Hourly wind speed at 10 m above the surface:											
RMSE	< 2 m s^{-1}	1.4	1.1	1.3	1.1	1.3	1.2	1.1	1.2	1.2	1.2
BIAS	< ± 0.5 m s^{-1}	0.6	−0.1	0.2	−0.1	0.0	0.1	−0.1	−0.1	0.0	−0.3
IOA	≥ 0.6	0.68	0.71	0.69	0.72	0.63	0.75	0.73	0.70	0.73	0.65
Hourly solar radiation at surface:											
R^2	1.00*	0.79	0.80	0.73	0.78	0.69	0.73	0.79	0.69	0.72	0.77
a	1.00*	1.11	1.11	1.16	1.15	1.11	1.13	1.11	1.07	1.08	1.11
b	0.00*	48.34	41.48	83.23	52.49	82.19	79.52	46.89	66.57	63.07	54.46

*Values for perfect fitting.

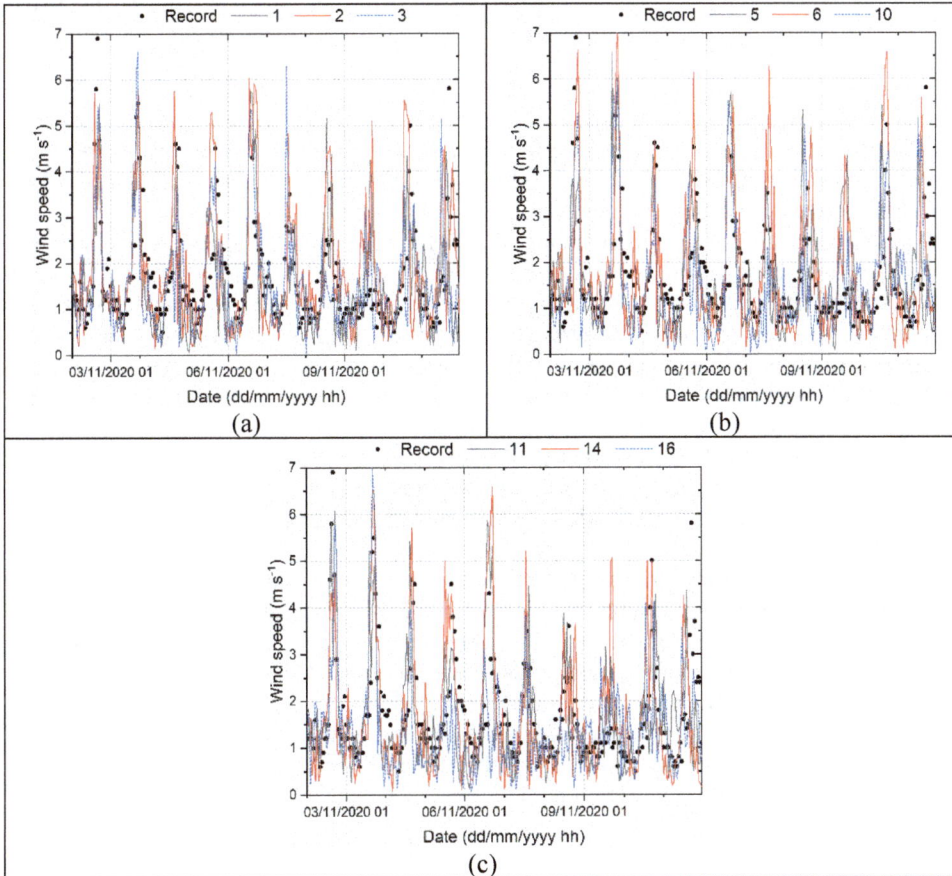

Figure 6:   Observed and modeled wind speed values at 10 m above the surface, with different parameterization (cu_physics options: 1, 2, 3, 5, 6, 10, 11, 14, 16).

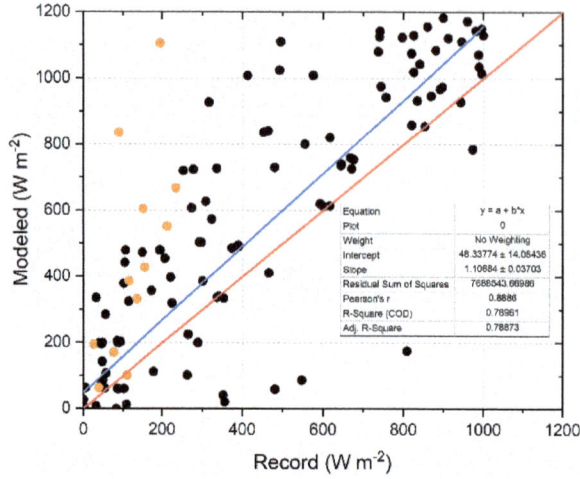

Figure 7:   Observed and modeled (cu_physics = 0) solar radiation values at the surface. The red line corresponds to a perfect linear fit. The blue line indicates the linear fitting. The orange dots indicate the pairs of 9 November 2020.

modeled solar radiation and wind speed values at 13:00 LT with the cumulus option 10 were 696 W m^{-2} and 2.8 m s^{-1}, corresponding to the moderately unstable category B. Therefore, if these computed parameters were used for forecasting purposes, the computed PM$_{2.5}$ concentrations would be lower than records.

Models using spatial resolution smaller than 4 km are generally considered convection-permitting, and they do not rely on cumulus parameterization schemes [15], therefore, following the assumption that the model is able to resolve organized convection at this resolution. This assumption is congruent with Zhang et al. [19], who reported a clear improvement on modeling precipitation in the Central Great Plains, in the eastern Kansas and western Missouri region (USA), without using a cumulus scheme at a 4 km resolution. Even, Zhang et al. [19] concluded that the advantage of modeling without a cumulus scheme should become more evident with finer resolutions such as 1 km. However, our results indicated that modeling with a spatial resolution of 1 km requires dedicated assessment to verify the usefulness of cumulus parameterization for regions with a complex topography and land-use heterogeneity.

Because most parameterized effects, such as the boundary layer, convection, microphysics, and surface schemes, are closely linked, it is necessary to study the influence of these components in the future. Also, it is necessary to assess the potential benefit of data assimilation and even the development of dedicated schemes for numerical simulation in the equatorial zone of the Andes and model meteorology and air quality with feedback.

Wind direction is another meteorological parameter directly involved in the dispersion of air pollutants. Although this parameter has been acceptably modeled in Cuenca, e.g. [4], [7], it was not evaluated in this contribution and should be included in future studies.

4 CONCLUSIONS

None of the cumulus parameterizations assessed in this contribution adequately modeled the meteorological variables on 9 November 2020, the day the lowest values of solar radiation, temperature, and wind speed were observed from 2 to 11 November 2020. It suggests that none of the cumulus options will provide proper values for days with significant drops in the magnitude of these parameters for forecasting purposes. If the modeled meteorological variables were used with a proper emission inventory for air quality forecasting, this system would provide proper air quality concentrations, except on days with drops in meteorological parameters similar to the observed on 9 November 2020.

Results indicated that modeling with a spatial resolution of 1 km for regions with a complex topography and land-use heterogeneity requires dedicated assessment to verify the usefulness of cumulus parameterization. It should not be assumed that for simulations with this spatial resolution, by default, it is better to turn off the convective cumulus option.

ACKNOWLEDGEMENTS

This research is part of the project "Emisiones y Contaminación Atmosférica en el Ecuador 2021–2022" and was funded by the USFQ Poli-Grants 2021 – 2022. Simulations were done at the High Performance Computing system at the USFQ.

REFERENCES

[1]   Baklanov, A. et al., Online coupled regional meteorology chemistry models in Europe: Current status and prospects. *Atmos. Chem. Phys.*, **14**(1), pp. 317–398, 2014.

[2]   Kumar, R., Peuch, V.-H., Crawford, J.H. & Brasseur, G., Five steps to improve air-quality forecasts. *Nature*, **561**(7721), pp. 27–29, 2018.

[3]   EMOV EP, Inventario de Emisiones Atmosféricas del Cantón Cuenca 2014, p. 88, 2016.

[4]   Parra, R. & Espinoza, C., Insights for air quality management from modeling and record studies in Cuenca, Ecuador. *Atmosphere*, **11**(9), p. 998, 2020.

[5]   EOSDIS Worldview, https://worldview.earthdata.nasa.gov/. Accessed on: 27 Feb. 2022.

[6]   Chow, F., Schär, C., Ban, N., Lundquist, K., Schlemmer, L. & Shi, X., Crossing multiple gray zones in the transition from mesoscale to microscale simulation over complex terrain. *Atmosphere*, **10**(5), p. 274, 2019.

[7]   Parra, R., Performance studies of planetary boundary layer schemes in WRF-Chem for the Andean region of southern Ecuador. *Atmospheric Pollution Research*, **9**(3), pp. 411–428, 2018.

[8]   Parra, R., Effects of aerosols feedbacks in modeling meteorology and air quality in the Andean region of southern Ecuador. Santiago de Compostela, Spain, pp. 39–50, 2021. http://library.witpress.com/viewpaper.asp?pcode=AIR21-004-1. Accessed on: 28 Feb. 2022.

[9]   Met Office, Grey zone project for model parameterizations. https://www.metoffice.gov.uk/research/approach/collaboration/grey-zone-project/index. Accessed on: 28 Feb. 2022.

[10] Kealy, J.C., Probing the "grey zone" of NWP: Is higher resolution always better? *Weather*, **74**(7), pp. 246–249, 2019.

[11] WRF, https://www2.mmm.ucar.edu/wrf/users/download/get_sources.html. Accessed on: 28 Jun. 2021.

[12] Research Data Archive, Commerce NC for EPWSSD of. NCEP FNL operational model global tropospheric analyses, continuing from July 1999. UCAR/NCAR, 2000. https://rda.ucar.edu/datasets/ds083.2/. Accessed on: 14 May 2021.

[13] WRF Users' Guide, https://www2.mmm.ucar.edu/wrf/users/docs/user_guide_v4/v4.0/contents.html. Accessed on 27 Feb. 2022.

[14] European Environment Agency, The application of models under the European Union's Air Quality Directive: A technical reference guide. LU Publications Office, 2011. DOI: 10.2800/80600.

[15] Berg, L.K., Gustafson, W.I., Kassianov, E.I. & Deng, L., Evaluation of a modified scheme for shallow convection: Implementation of CuP and case studies. *Monthly Weather Review*, **141**(1), pp. 134–147, 2013.

[16] Grell, G.A. & Freitas, S.R., A scale and aerosol aware stochastic convective parameterization for weather and air quality modeling. *Atmos. Chem. Phys.*, **14**(10), pp. 5233–5250, 2014.

[17] Zhang, C., Wang, Y. & Hamilton, K., Improved representation of boundary layer clouds over the southeast Pacific in ARW-WRF using a modified Tiedtke cumulus parameterization scheme. *Monthly Weather Review*, **139**(11), pp. 3489–3513, 2011.

[18] De Visscher, A., *Air Dispersion Modeling: Foundations and Applications*, Wiley: USA, 634 pp., 2014. https://www.wiley.com/en-us/Air+Dispersion+Modeling%3A+Foundations+and+Applications-p-9781118078594. Accessed on: 4 Mar. 2022.

[19] Zhang, Y., Roundy, J.K. & Santanello, J.A., Evaluating the impact of model resolutions and cumulus parameterization on precipitation in NU-WRF: A case study in the Central Great Plains. *Environmental Modelling and Software*, **145**, 105184, 2021.

NUTRIENTS IN MARGINAL LAND SOILS AND THEIR POTENTIAL EFFECT ON THE ENVIRONMENT

NICOLE RODRIGUEZ, TIMOTHY HO, ZHONGWEI SHI & JULIA LU
Department of Chemistry and Biology, Ryerson University, Canada

ABSTRACT

Increasing global population leads to an increase in demand for foods and cleaner energy such as biofuel and bioenergy that are produced from feedstocks. Utilizing marginal land for production of these feedstocks alleviates the competition of fuel versus food that comes with use of prime agricultural land. Canada has a large area of marginal land. Sorghum is an important plant for food, fodder, and forage production. It is regarded as a nature-cared plant with low input requirements and is recommended as a top crop for removing carbon from the atmosphere. As a part of a collaborative project to develop a system for producing biomass (sorghum) on marginal land in Canada, this research focuses on the species and their distribution, mobility and availability (to plants) of nitrogen (N) and phosphorous (P) in marginal land soils from selected locations in Canada. US EPA method 1312 was followed to simulate the leaching process of nutrients from soils in the natural environment. Colorimetry and ICP-OES were used for the determination of the nutrient species. Preliminary results show that the predominant leachable and plant-usable form of nitrogen is nitrate (NO_3^-) while the majority of phosphorus in the soil is not water leachable; depth variation of leachable nitrogen and phosphorus species in the soils is indicated; the concentrations of nitrate in the soils increased shortly after N-fertilizer application but the level decreased to that observed before planting, suggesting that atmospheric precipitate/deposition can move nitrogen from marginal land soils to surface water.

Keywords: soil nutrients, biomass, marginal land, mobility, nitrogen, phosphorus.

1 INTRODUCTION

1.1 Renewable energy

Renewable energies are gaining attention and importance recently with the awareness of limited non-renewable energies and their impact on the environment. Biomass has the potential to be used in a variety of bioenergy applications. There is evidence to predict that global bioenergy could evolve to meet the global energy demand by 2050 [1], [2]. The production of such feedstocks is, however, depending on the land and its cost. The major feedstocks for biofuel consist of agricultural crops such as corn, sugarcane, and soybean and their production compete land for food production. There has been a shift in activity towards growing alternative biofuel feedstocks (e.g., sorghum) and utilizing non-agricultural (e.g., marginal) land [3].

1.2 Marginal land

Marginal land is broadly defined as land that is either not economically suitable for agricultural use, or less productive than agricultural land [4], [5]. The classification of marginal land varies, but common characteristics involve soil quality, climatic conditions and land attributes such as evaluation, irregular terrain. Canada has 9.48 million hectares of land that is classified by the Canada Land Inventory (CLI) as marginal land (Class 4–6) [6]. There is great potential for Canada to produce biomass on its marginal land [3]. Stoof et al. [5] noted concerns of nutrient availability for these marginal soils. Sorghum is regarded as a

WIT Transactions on Ecology and the Environment, Vol 259, © 2022 WIT Press
www.witpress.com, ISSN 1743-3541 (on-line)
doi:10.2495/AWP220021

nature-cared plant with low input requirements, and it is recommended as a top crop for removing carbon from the atmosphere [7], [8].

1.3 Nitrogen and phosphorus

Nitrogen and phosphorus are essential elements in plant growth and development. Nitrogen supply in a crop system is predominantly represented by the level of N-species available for plant use. Inorganic nitrate (NO_3^-) and ammonium (NH_4^+) are the main forms of nitrogen that can be taken by plants, thus be removed from the soils [9]. In addition, ammonium (NH_4^+) can move from surface soil to the atmosphere through volatilization; NO_3^- can leave soil through denitrification process and it can be easily removed by leaching or runoff. Phosphorus is never found in high concentrations, making it highly conserved and is stored through sedimentation and rock formation [10], [11]. Phosphorus may be bound to metal oxides such as aluminium or iron oxides, which are much harder to leach out due to the common ion effect governed by small solubility product constants [12], [13]. Organic phosphorus can be categorized into different organic phosphate compounds in soil. Both inorganic and organic phosphorus are non-labile when in a solid-state and they make up most of the soil phosphorus. The most soluble form of inorganic phosphorus is orthophosphate (PO_4^{-3}) and is can easily be removed by leaching and runoff if not taken up by plants or converted into a non-labile form [14].

1.4 Water pollution

Fertilizer is commonly used to promote crop production. Many factors, such as placement, timing, type of fertilizer, and rate of application, have to be considered in utilizing fertilizer. Rate of fertilization is of particular importance, as excess fertilizer to a crop system can be easily lost via various processes [9], [15]. Excessive use of fertilizers has caused negative impacts through runoff which pollutes water and surrounding ecosystems [16].

1.5 Research objectives

This research studies the species of nutrients (N and P) in soil at different depths through the growing period of sorghum on marginal land and assesses the potential effect of these species on water quality.

2 EXPERIMENTAL

Field studies were carried out in 2019. Three sorghum hybrids: CSSH45, 10AX118, and 10AX131, were tested. Half of the plots at each field received one application of N-fertilizer (urea, 46-0-0, rate of 45 kg ha^{-1}) about two weeks after planting.

2.1 Study sites

Three field sites in Ontario, Canada are evaluated in this research and they are located in London (43°01'49.0"N, 81°12'23.6"W), Simcoe (42°51'35.0"N, 80°16'17.0"W), and Ottawa (45°19'30.0"N, 75°52'55.2"W). The field designs are shown in Fig. 1.

2.2 Sample collection and preparation

Soil cores were collected from each site three times during the growing period: before planting, about two weeks after the fertilizer application (after planting) and at harvest. The

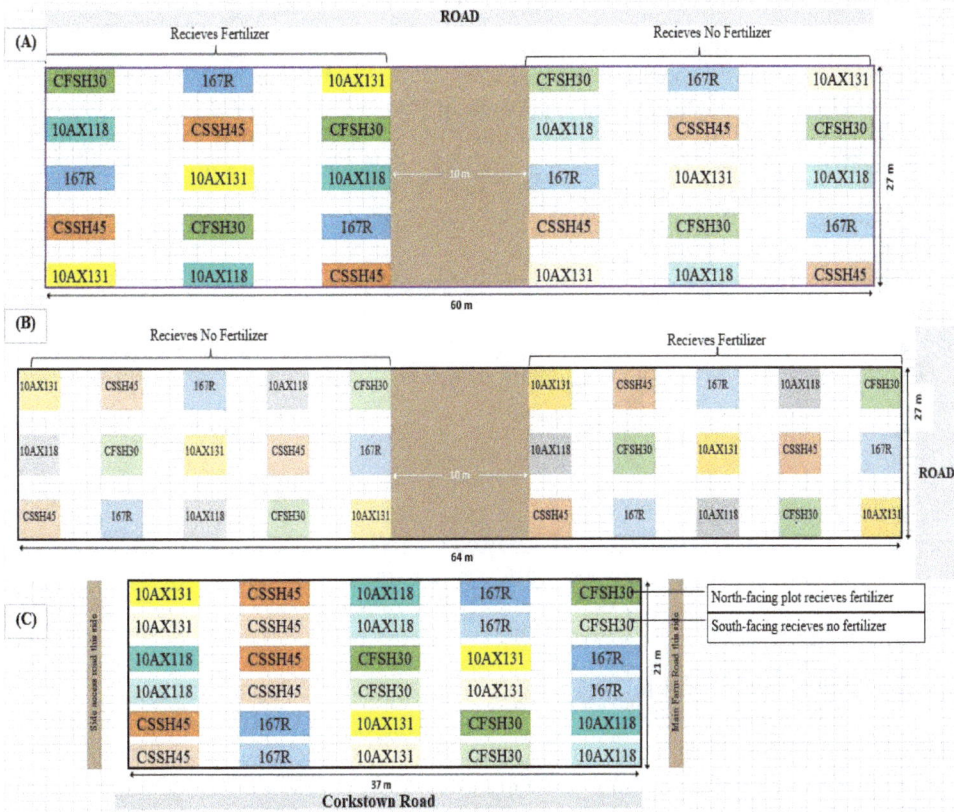

Figure 1: Field plot design for (A) London; (B) Simcoe; and (C) Ottawa, Canada.

soil samples at 0–10 cm and 20–30 cm depths from the soil cores were collected. After the collection, soil samples were stored in sealed Ziploc bags and stored in a freezer until analysis. Field soil samples were thawed for 24 hours, grinded, and passed through a 2 mm sieve using a mortar. The homogenized soil samples were extracted using water of pH 5.0 following EPA method 1312 [17] at a mixing ratio of 1.5 g soil:30 g water (1:20) for 18 ± 2 hours. Centrifugation was used to separate the liquid (soil extract) from the soil residuals.

2.3 Analysis methods

2.3.1 Nitrogen species

Concentration of NO_3^- ([NO_3^-]), NO_2^- ([NO_2^-]), and NH_4^+ ([NH_4^+]) in the soil extracts were determined (2.00 g soil extract for NO_3^- and NO_2^-, 3.00 g soil extract for NH_4). The [NO_2^-] was analyzed using Griess reaction, a scaled-up version outlined by Hood-Nowotny et al [18]. NO_3^- was reduced to NO_2^- with vanadium chloride (VCl_3) then determined using the same Griess reaction as for NO_2^- [19]. After the colour reaction, the [NO_2^-] was determined using UV-Vis (Cary 60, Agilent) at 540 nm and [NO_3^-] was then calculated. The [NH_4^+] is analyzed using a modified Berthelot reaction catalyzed by sodium nitroprusside [20] and is then determined using UV-Vis (Cary 60 Agilent) at 687 nm. The analysis was done in triplicate.

2.3.2 Phosphorus

Orthophosphate concentration ($[PO_4^{-3}]$) and total phosphorous ($[P]_{total}$) in soil extracts were determined (1.00 g soil extract for PO_4^{-3}, 0.35 g soil for total phosphorous). The analysis of PO_4^{-3} based on molybdenum blue reaction [21] followed by UV-Vis (Cary 60, Agilent) determination at 890 nm. Microwave digestion (EPA 3051A) + ICP-OES were used for the determination of $[P]_{total}$ in the soil samples [22]. The analysis was done in triplicate.

2.4 Data QA/QC

The analysis methods were validated using standard reference materials (nutrients in soil; lot: LRAC0708, and anions in soil; lot: LRAC1295, Sigma Aldrich). One or more blank solutions and a standard solution measurement were carried out each day of analysis to monitor changes or trends in each method. The limit of detection (LOD) for each species is defined as:

$$LOD = \frac{3 * S_{blank}}{b},$$

where S_{blank} is the standard deviation of the blank values, b is the slope of the calibration curve [23]. The LODs were 0.03, 0.01, and 0.04 mg kg^{-1} for NO_3^-, NO_2^-, and NH_4^+. The LODs were 0.40 mg kg^{-1} for PO_4^{-3} and 2.40 mg kg^{-1} for total P.

2.5 Data analysis

One-way ANOVA was used for data analysis to determine whether or not the difference in nitrogen levels between the groups is significant. The groups evaluated were between the top and bottom soil layers within each field, plots with and without fertilizer application within each field, fields between agricultural and marginal lands, temporal effects within each field over the growing season.

3 RESULTS AND DISCUSSION

3.1 Nitrogen species

A total of 453 soil samples were analyzed and the results show $[NO_3^-]$ in 98% of the samples, $[NO_2^-]$ in only 34%, and $[NH_4^+]$ in 8% of the samples above their LODs. NO_2^- is an intermediate species in the nitrification reaction, which rarely accumulates in soil. For NH_4^+, as it is a positive cation, it tends to be more tightly bound to soil particles which are inherently negatively charged. Since acidified water was used as the extracting solution instead of salt solution, it was not expected the $[NH_4^+]_{soil}$ to be high. The following results and data analysis will, therefore, focus on NO_3^- in the soil samples only.

The NO_3^- concentrations in the soil samples collected before planting at 0–10 cm and 20–30 cm depths from all three sites are shown in Fig. 2. At each site, the $[NO_3^-]$ is higher in the top layer soil but the values in each soil layer are not statistically significant (p-values ≥ 0.05) at all three sites, regardless soil type (agricultural vs marginal land).

Fig. 3 shows the $[NO_3^-]$ in the soil samples collected about two weeks after fertilizer application and at harvest from the plots with different sorghum hybrids and with and without the application of the N-fertilizer from all the experimental sites. The results show similar trends across all three hybrids in terms of $[NO_3^-]$ within each field. There is a general decrease in $[NO_3^-]$ at deeper soil layer from both the unfertilized and fertilized plots. Statistical

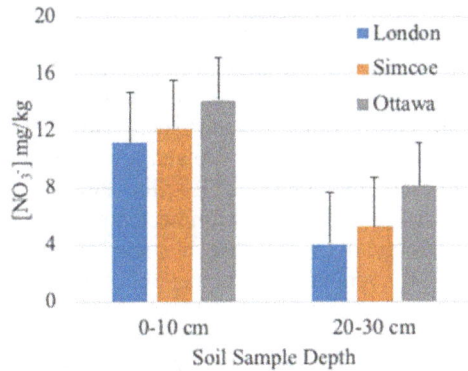

Figure 2: $[NO_3^-]$ in the soil of two layers from the field sites before planting (n = 4). Whiskers represent standard errors.

analysis resulted in p-values < 0.05 thus the difference was significant for the $[NO_3^-]$ values between the two soil layers within each hybrid at all three sites during the growing period. This can be explained that soil was tilled at about 15 cm deep as agricultural practice, whereas the untilled soil located below was more compact. This physical difference could limit NO_3^- downward movement with water as the pore space is limited. After planting $[NO_3^-]$ is higher in the soil from the plots with urea applied, especially at the 0–10 cm as urea was directly applied on the surface. However, the difference is less apparent at harvest. Further data analysis will therefore be focused on 0–10cm soil layer only.

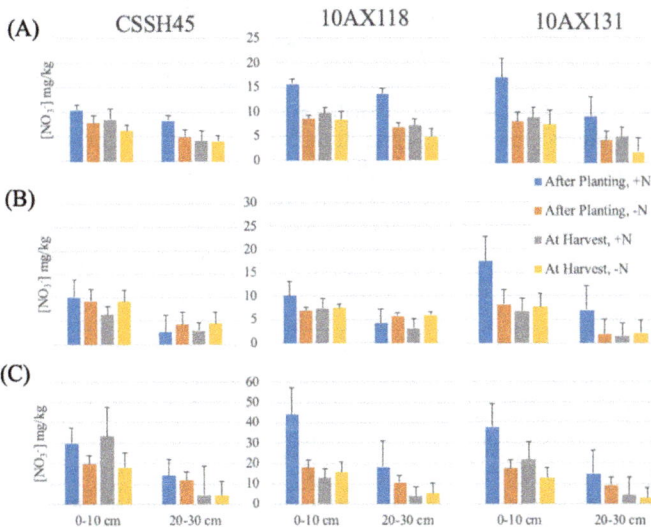

Figure 3: $[NO_3^-]$ levels at two soil depths from the plots with sorghum hybrids (CSSH45, 10AX118, 10AX131) at (A) London; (B) Simcoe; and (C) Ottawa fields. –N = unfertilized plots; +N = fertilized plots (n = 3). Whiskers represent standard errors.

Statistical analysis (Table 1), shows that N-fertilizer application has significant effect on the $[NO_3^-]_{soil}$ levels after planting at the London and Ottawa sites. This difference was expected as the application of urea two weeks before the sample collection would theoretically increase the amount of N available in soil. The same effect, however, is not indicated for the Simcoe site, possibly due to factors such as lower organic content, compared with that in the agricultural soil at the London site and lower $[NO_3^-]$, compared with that at the Ottawa site, in the soil before planting (Fig. 2), soil texture, etc. At harvest, however, the effect is no longer significant at all three sites as indicated in Table 1. The results in Table 1 also show that sorghum variety does not have a statistically significant impact on the $[NO_3^-]_{soil}$. The same finding was reported by Chen et al. for a marginal site in China [24].

Table 1:   Summary of statistical analysis for the effects of N-fertilizer (N) and sorghum hybrid (H) on $[NO_3^-]$ in the top 10 cm soil collected after planting and at harvest from the field sites.

Collection period	Field	N	H
After planting	London	*	ns
	Simcoe	ns	ns
	Ottawa	*	ns
At harvest	London	ns	ns
	Simcoe	ns	ns
	Ottawa	ns	ns

* Denotes a significant effect at p-value < 0.05; ns denotes non-significant effects, p-value ≥ 0.05.

Since sorghum hybrids have no significant effect on the $[NO_3^-]$ in the soils, the data of all hybrids were combined at each collection to assess the temporal effect on the $[NO_3^-]$ in the soil throughout the growing season, noting that three collection periods (i.e., before planting, after planting, at harvest) for unfertilized while only two (i.e., after planting and at harvest) for fertilized condition were used for the assessment. Statistical analysis results summarized in Table 2 show that, throughout the growing season, the difference in the $[NO_3^-]$ in the unfertilized soil is significant at the agricultural site in London but not significant at the marginal sites in Simcoe and Ottawa, but the difference in the $[NO_3^-]$ in the N-fertilized soil is significant at all the sites tested, regardless soil type.

Table 2:   Summary of statistical analysis for temporal effect on the $[NO_3^-]_{soil}$ in the top 10 cm soil from the field sites over the growing season.

Field	Soil depth	Temporal effect on unfertilized samples	Temporal effect on fertilized samples
London	0–10 cm	*	*
Simcoe	0–10 cm	ns	*
Ottawa	0–10 cm	ns	*

* Denotes a significant effect at p-value < 0.05; ns denotes non-significant effects, p-value ≥ 0.05.

To further evaluate the changes in the $[NO_3^-]_{soil}$ over the 2019 growing season, NO_3^- in the top 10 cm soil (kg ha^{-1}) during the growing period at each site was calculated and the values are listed in Table 3. An average bulk density of 1.45 g cm^{-3} for sandy loam soils was used for the estimate [25]. The 20–30 cm depth was excluded from this comparison as the soil at this depth pasts the tillage depth.

Table 3: Summary of $[NO_3^-]$ (kg ha^{-1}) in the top 10 cm soil at the collection periods from the field sites.

Fertilizer / Field	Before planting	After fertilizer application (theoretical)	After planting		At harvest	
	$-N$	$+N$	$-N$	$+N$	$-N$	$+N$
London	14.5 ± 3.7	59.5	11.8 ± 2.2	19.5 ± 5.7	9.7 ± 3.0	12.1 ± 2.3
Simcoe	15.5 ± 5.4	60.5	9.8 ± 3.8	15.0 ± 9.8	9.9 ± 3.9	9.9 ± 2.3
Ottawa	19.3 ± 3.1	64.3	21.9 ± 6.5	37.6 ± 21.0	16.0 ± 10.1	21.4 ± 17.0

The values in Table 3 show that, overall, the NO_3^- in the soils with and without N-fertilizer application decreased from before planting to at harvest, but the changes are not statistically significant in the soils from the plots without N-fertilizer application at all sites tested.

Nitrogen in the soils with N-fertilizer application decreased from after fertilizer application to at harvest at each field site, with net change (i.e., N-loss) of 47.4, 50.6, and 42.9 kg ha^{-1} in the soil for London, Simcoe and Ottawa site, respectively. NO_3^- can be removed from soil by various processes such as plant uptake, denitrification process, leaching, runoff. The results obtained by Tian's group (Agriculture and Agri-Food Canada, personal communication) show that N-fertilizer application had no significant effect on the dry sorghum mass produced at each of the sites tested. Data analysis showed that $[NO_3^-]_{soil}$ was anti-correlated to the amount of precipitate during the collection periods after N-fertilizer application and at harvest, suggesting leaching and runoff play a role in removing $[NO_3^-]$ from the soils.

3.2 Phosphorus

The $[PO_4^{-3}]$ in 88% of the soil samples analyzed were above LOD. The $[PO_4^{-3}]$ in the soil samples collected before planting from all the field sites are shown in Fig. 4. It can be seen that the $[PO_4^{-3}]$ levels in the 0–10 cm soil layer are higher than those in the 20–30 cm soil layer. Data analysis resulted in p-values < 0.05, confirming the difference is statistically significant.

The $[PO_4^{-3}]$ in the soil samples collected after planting and at harvest from all the field sites are shown in Fig. 5 and statistical results are summarized in Table 4. The $[PO_4^{-3}]$s in all the soil samples with N-fertilizer applied between the two soil layers are statistical different for both after planting and at harvest, but the values in the two soil layers without the fertilizer application are not statistically significantly different, except in the soils from plots with 10AX118 sorghum hybrid after planting. The $[PO_4^{-3}]$ in the soils at both depths at Ottawa site is relatively low, but statistical analysis suggests that difference is significant between the depths.

Total P in the soil samples, ranged from 479 to 938 mg kg^{-1} for the London site; 346 to 1106 mg kg^{-1} for the Simcoe site and 386 to 1156 mg kg^{-1} for the Ottawa site. Statistical analysis showed no significant difference on the $[P]_{total}$ for all except the London field site. $[P]_{total}$ profile pattern is observed, more prominently, in the soils at the London site (i.e.,

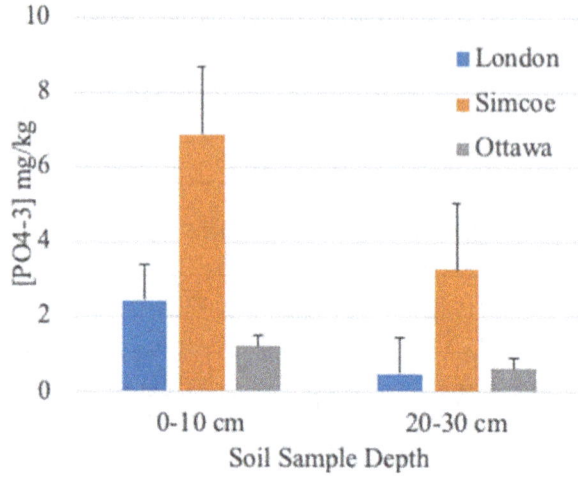

Figure 4:   Comparison of $[PO_4^{-3}]$ levels in two soil depths at three fields before planting (n = 4). Whiskers represent standard errors.

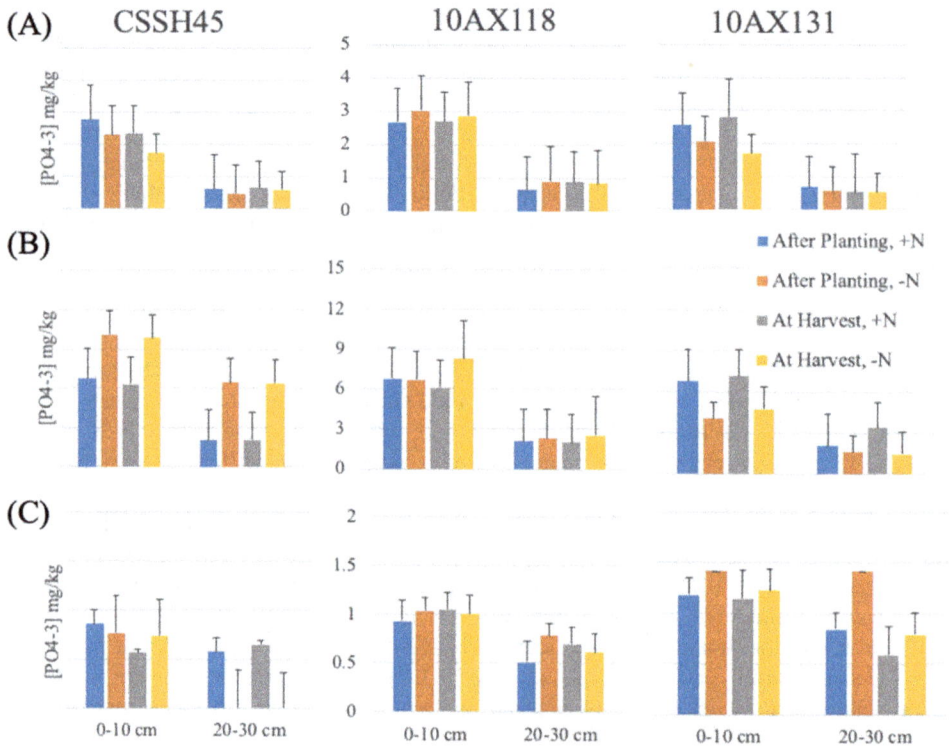

Figure 5: $[PO_4^{-3}]$ levels (n = 3) at two soil depths with three different sorghum hybrids (CSSH45, 10AX118, 10AX131) at (A) London; (B) Simcoe; and (C) Ottawa field sites, 2019. –N = unfertilized plots; +N = fertilized plots. Missing bars are due to values <LOD. Whiskers represent standard errors.

Table 4:   Summary of statistical analysis for the effects of nitrogen fertilizer (N) and sorghum hybrid (H) on $[PO_4^{-3}]$ at the corresponding depth and collection periods at the field sites.

		After planting		At harvest	
Field	Soil depth	N	H	N	H
London	0–10 cm	ns	ns	ns	ns
	20–30 cm	ns	ns	ns	ns
Simcoe	0–10 cm	ns	ns	ns	ns
	20–30 cm	ns	ns	ns	ns
Ottawa	0–10 cm	ns	ns	ns	ns
	20–30 cm	ns	ns	ns	ns

ns denotes non-significant effects, p-value ≥ 0.05.

agricultural land), likely due to the past application of P fertilizer. Ottawa did not exhibit the same pattern: $[P]_{total}$ is highly homogenous across the two depths in Ottawa regardless of sorghum hybrids or the N-fertilizer application. The collection periods appeared to have no statistically significant impact on $[P]_{total}$ in the soils. The results show that less than 1% P in the soils is present as PO_4^{-3}.

Table 4 shows the summary of statistical analysis that evaluates the effect of N-fertilizer and sorghum hybrid on $[PO_4^{-3}]_{soil}$, showing that neither N-fertilizer nor sorghum hybrid had statistically significant effect on the $[PO_4^{-3}]$ in the soils.

4 CONCLUSION

Analysis of nitrogen species showed that NO_3^- is the predominant leachable and mobile N-species in the soils. N-fertilizer increased the $[NO_3^-]$ in the soils but the level decreased to that observed in the soils collected before planting. Leach and runoff play roles in removing NO_3^- from the soil regardless the land types and fertilizer acts as a source of pollutants to freshwater. The majority of P in soils is not mobile thus will unlikely have significant effect on water quality under the experimental conditions.

ACKNOWLEDGEMENT

Financial support from Biomass Canada Cluster (BMC) through Agriculture and Agri-Food Canada's AgriScience program and industry partners to Julia Lu.

REFERENCES

[1]   Ladanai, S. & Vinterbäck, J., *Global Potential of Sustainable Biomass for Energy*, SLU, Institutionen för Energi Och, p. 29, 2009.

[2]   World Bioenergy Association, Global Bioenergy Statistics, 2019.

[3]   Liu, T. et al., Bioenergy production on marginal land in Canada: Potential, economic feasibility, and greenhouse gas emissions impacts. *Appl. Energy*, **205**, pp. 477–485, 2017.

[4]   James, L.K., Theory and identification of marginal land and factors determining land use change. *Science*, pp. 1–99, 2010.

[5]   Stoof, C.R. et al., Untapped potential: Opportunities and challenges for sustainable bioenergy production from marginal lands in the northeast USA. *Bioenergy Res.*, **8**, pp. 482–501, 2015.

[6]   Government of Canada, Canada Land Inventory (CLI). https://sis.agr.gc.ca/cansis/nsdb/cli/classdesc.html.

[7]   Holou, R.A.Y. & Kindomihou, V.M., The biofuel crops in global warming challenge: Carbon capture by corn, sweet sorghum and switchgrass biomass grown for biofuel production in the USA. *Front. Bioenergy Biofuels*, 2017.

[8]   Nenciu, F., Vlăduț, V., Nae, G., Popa, L.-D. & Constantin, O.E., Developing efficient practices for producing bioethanol from sweet sorghum, evaluating several varieties of the plant and growing environments. *E3S Web Conf.*, **286**, 03020, 2021,

[9]   Jämtgård, S., Näsholm, T. & Huss-Danell, K., Nitrogen compounds in soil solutions of agricultural land. *Soil Biol. Biochem.*, **42**, pp. 2325–2330, 2010.

[10] Wood, T., Bormann, F.H. & Voigt, G.K., Phosphorus cycling in a northern hardwood forest. *Biological and Chemical. American Association for the Advancement of Science*, **223**(4634), pp. 391–393, 2008.

[11] Perring, M.P., Hedin, L.O., Levin, S.A., McGroddy, M. & De Mazancourt, C., Increased plant growth from nitrogen addition should conserve phosphorus in terrestrial ecosystems. *Proceedings of the National Academy of Sciences of the United States of America*, **105**(6), pp. 1971–1976, 2008.

[12] Lide, D.R., Solubility product constants, *CRC Handbook of Chemistry and Physics*, 2006.

[13] Fink, J.R., Inda, A.V., Tiecher, T. & Barrón, V., Iron oxides and organic matter on soil phosphorus availability. *Ciencia e Agrotecnologia*, pp. 369–379, 2016.

[14] Frossard, E., Condron, L.M., Oberson, A., Sinaj, S. & Fardeau, J.C., Processes governing phosphorus availability in temperate soils. *Journal of Environmental Quality*, **29**(1), pp. 15–23. 2000.

[15] Kim, J. & Rees, D.C., Nitrogenase and biological nitrogen fixation. *Biochemistry*, **33**, pp. 389–397, 1994.

[16] Sharpley, A.N., Chapra, S.C., Wedepohl, R., Sims, J.T., Daniel, T.C. & Reddy, K.R., Managing agricultural phosphorus for protection of surface waters: Issues and options. *Journal of Environmental Quality*, 1994.

[17] United States Environmental Protection Agency (USEPA), Method 1312: Synthetic precipitation leaching procedure, Vol. 1, 1994.

[18] Hood-Nowotny, R., Umana, N.H.-N., Inselbacher, E., Oswald-Lachouani, P. & Wanek, W., Alternative methods for measuring inorganic, organic, and total dissolved nitrogen in soil. *Soil Sci. Soc. Am. J.*, **74**, pp. 1018–1027, 2010.

[19] Miranda, K.M., Espey, M.G. & Wink, D.A., A rapid, simple spectrophotometric method for simultaneous detection of nitrate and nitrite. *Nitric Oxide – Biol. Chem.*, **5**, pp. 62–71, 2001.

[20] Weatherburn, M.W., Phenol-hypochlorite reaction for determination of ammonia. *Anal. Chem.*, **39**, pp. 971–974, 1967.

[21] Harwood J.E., van Steenderen R.A. & Kühn A.L., A rapid method for orthophosphate analysis at high concentrations in water. *Water Research*, **3**(6), pp. 417–423, 1969.

[22] United States Environmental Protection Agency (USEPA), Method 3051A: Microwave assisted acid digestion of sediments, sludges, soils, and oils, 2007.

[23] Skoog D.A., West D.M., Holler F.J. & Crouch S.R., Fundamentals of analytical chemistry. Cengage Learning, p. 214, 2014.

[24] Chen, F. et al., Effects of nitrogen fertilization on soil nitrogen for energy sorghum on marginal land in China. *Agron. J.*, **109**, pp. 636–645, 2017.

[25] Government of Canada, Canadian Soil Information Service (CanSIS). https://sis.agr.gc.ca/cansis/index.html.

ENSEMBLE DEEP LEARNING FOR CLASSIFICATION OF POLLUTION PEAKS

PHUONG N. CHAU[1], RASA ZALAKEVICIUTE[2] & YVES RYBARCZYK[1]
[1]School of Information and Engineering, Dalarna University, Sweden
[2]Grupo de Biodiversidad Medio Ambiente y Salud (BIOMAS), Universidad de Las Américas, Ecuador

ABSTRACT

The concentration peaks of atmospheric pollutants are the most challenging and important phenomena in air quality forecasting. The fact that these elevated levels of pollution do not seem to follow any specific pattern explains why current models still struggle to provide an accurate prediction of these harmful events for human health. The present study tackles this issue by testing several supervised learning methods to discriminate between peak and no peak of concentrations of five contaminants: NO_2, CO, SO_2, $PM_{2.5}$, and O_3. The classification performance of ensemble decision tree (gradient boosting machine (GBM)) models and ensemble deep learning (EDL) models are compared. The results reveal that the EDL outperforms the GBM model. An analysis of the variable importance (SHapley additive exPlanations (SHAP)) shows that both temporal and meteorological features have an impact on the proposed models. In particular, time of day and wind speed are the most important features to explain the performance of the ensemble DL models.
Keywords: machine learning, deep learning, air pollution forecasting, data-driven modelling.

1 INTRODUCTION

Traditionally, air quality modelling has been done using atmospheric chemical transport models (CTMs), which provide a powerful framework for describing emission patterns, meteorology, and chemical transformation processes [1]. More recently, machine learning (ML) modelling has demonstrated its efficiency and reliability in predicting pollutant concentrations in the atmosphere [2]–[4]. Moreover, Grange et al. used ML and meteorological parameters to develop weather normalized models (WNMs) for developing air contaminant prediction models [5]. Rybarczyk and Zalakeviciute [4], Barré et al. [2] and Ceballos-Santos et al. [3] developed WNMs using gradient boosting machine (GBM) to simulate and quantify the effects of human activities on the environment in the context of COVID-19 lockdowns.

The past years have seen the rapid developments of deep learning (DL) models that have been applied to both timeseries data and air quality modelling. Particularly, long-short term memory (LSTM) model was developed in 1997 by Hochreiter and Schmidhuber [6] to capture both long- and short-term memory from the sequence data. Afterwards, LSTM models have been more adapted for time series prediction [7], [8] and have also been applied for predicting CO, NO_2, O_3, PM_{10}, SO_2 and pollen concentrations in Madrid [9]. In 2019, Krishan et al. used LSTM to develop air quality models for India [10]. After recognizing the limitations of LSTM, Schuster and Paliwal combine two hidden LSTM layers in the opposite direction to create the bidirectional recurrent neural network (BiRNN) [11]. It has been demonstrated to be an effective DL architecture for sequence data with two hidden LSTM layers in opposite directions. In 2019, BiRNN was used by Li et al. [12] to capture deeper characteristics of timeseries data. Besides, Tong et al. used BiRNN for modelling the $PM_{2.5}$ concentrations and exploring the correlations of spatiotemporal properties [13]. Also, Zhang et al. developed a hybrid model based on BiRNN for $PM_{2.5}$ forecasting, which outperformed the traditional models [14].

WIT Transactions on Ecology and the Environment, Vol 259, © 2022 WIT Press
www.witpress.com, ISSN 1743-3541 (on-line)
doi:10.2495/AWP220031

Although the performance of air quality forecasting is improving, the concentration peaks seem to be extremely difficult to handle or predict [15]–[17]. The reason is that elevated levels of pollutants do not seem to follow any specific pattern. Consequently, the current models continue to struggle to provide an accurate prediction of harmful air quality. In addition, when we can determine the concentration peak, the other concentrations are equal or lower than the peak value. Therefore, the air pollutant trend is not going up after the highest concentration, and we can implement many strategies to improve the predicted models such as external features. A model able to detect peaks can provide early warnings of air pollutant concentrations exceeding the WHO guidelines [18].

This study addresses the daily peaks classification problem of five air pollutants (NO_2, SO_2, CO, O_3 and $PM_{2.5}$) by proposing advanced techniques, based on ensemble decision tree GBM and ensemble deep learning (EDL), to automatically classify peak vs no peak of pollutant concentrations. Additionally, we leveraged SHapley Additive exPlanations (SHAP) to explore the variable correlations in the EDL models. This method enables us to discover the principal factors responsible for the daily peak of concentrations.

The main goal of this study is to create supervised learning models that use data-driven techniques to classify the daily air pollutants according to the peaks of concentration. The remainder of this paper includes four sections. Section 2 describes the study site, data collection, and data processing. Section 3 depicts the supervised methods. The results and discussion are presented in Section 4. Finally, this research is summarized and concluded in the last section.

2 MATERIALS

2.1 Study site

Bellisario (2,835 meters above sea level (m.a.s.l), coord. 78°29'24" W, 0°10'48" S) study site is one of the monitoring stations in Quito located on a school and in a heavily populated area of the capital city of Ecuador. The weather in the study site is mild and consistent throughout the year. These characteristics are caused by the city's two distinct seasons. The wet season is from September to May, and the dry season is between June and August [19]. Furthermore, this city is a high elevation city established at 2,850 m.a.s.l. with a population of about 2.2 million people in 2011 [20], [21]. Due to 30% reduction of oxygen concentrations at this altitude, traffic emissions are the principal causes of long-term pollution problems for this urban center [19], [22].

2.2 Data

The data is provided by the Secretariat of the Environment of the Municipality of the Metropolitan District of Quito. All devices follow the Environmental Protection Agency of the United States (US-EPA) standards. The data is from a central urban study site Bellisario with five pollutants, namely NO_2, CO, SO_2, $PM_{2.5}$, O_3 and seven meteorological features such as solar radiation (SR), wind direction (WD), wind speed (WS), atmospheric pressure (p), precipitation (Prec), temperature (T) and relative humidity (RH). Additionally, we created four temporal features such as "Julian day" (or day of the year), "week day" (day of the week), hours (the time of the day), and index (i.e., starting from 1 January 2014 and increasing by one at each instance). According to research by Grange et al. [5] and Henneman et al. [23], the temporal variables are independent features in meteorological normalization models. "Julian day", "week day", and "hours" are indicators for emission pattern cycles rather than direct influences on levels of pollutants in the atmosphere. "hours", for example, is a term used to explain cyclic emissions such as traffic-related rush hour emissions. "week day" represents weekdays or weekends in a week. Meanwhile, "Julian day" is a seasonal term that strongly presents seasonal emissions or pollutant concentrations.

2.3 Pre-process data

We selected the data for a total of five years from 1 January 2014 to 31 December 2018. The instance is removed from the dataset if the missing value is pollutant concentrations. The missing values in meteorological features are replaced by their averages. Since the filled values in this phase are less than 1%, it allows us to use the average method for imputation [24], [25]. Next, the output features will be labelled with two labels: 0 and 1. The daily peak concentration is labelled "1", and the label "0" is randomly selected for another concentration on the same day. We chose the label "0" randomly, but it must differ from the highest daily concentration. It allows us to generate balanced datasets with two classes.

3 METHODS

Fig. 1 illustrates an overview of our process. First, we split the data into training and testing set with 80% (four years from 1 January 2014 to 31 December 2017) for training and 20% (one year from 1 January 2018 to 31 December 2018) for testing. Second, we develop EDL models based on LSTM and BiRNN architectures for classifying the peak concentrations with five air pollutants (NO_2, SO_2, CO, O_3 and $PM_{2.5}$) in Section 3.2. Additionally, we develop the GBM model for this problem in Section 3.1. Next, the GBM and EDL models are compared, which is based on the assessment metrics (F1-score and AUC score) to evaluate the classification performance (Section 3.3). The models with the highest F1 and AUC scores are selected as the best models for classification problems. According to the best models, we can use it to identify the daily peak values of air pollutants. Besides, it allows us to explore the variable correlations and discover the significant factors in the pollutant concentration peaks with SHAP in Section 3.4.

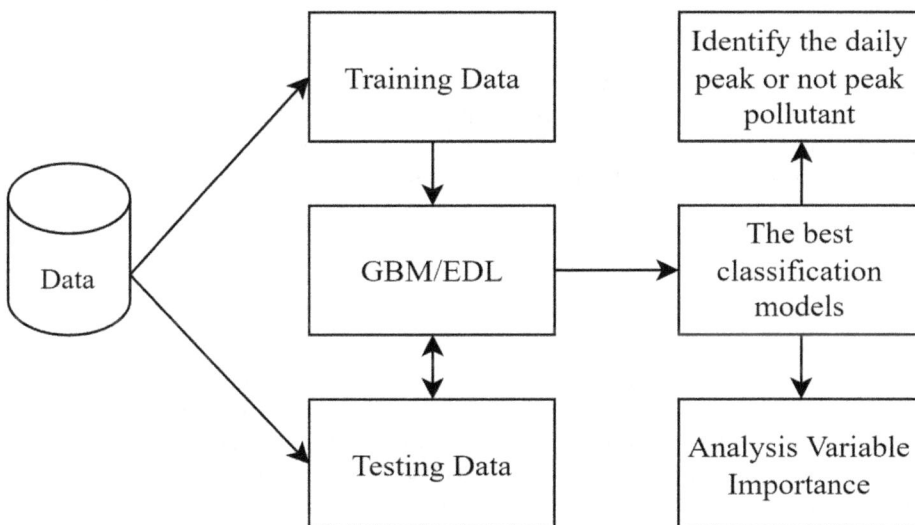

Figure 1: Workflow of our process.

3.1 Gradient boosting machine

Friedman proposed GBM in 2001 as a decision tree-based ensemble learning algorithm [26]. Typically, an ensemble model is developed to obtain a better generalization model from weak learning models. GBM algorithm builds regression trees for all features from the dataset sequentially, which means that the trees are built independently. When all trees are generated, the final output from this algorithm is obtained by eqn (1). $\hat{f}^n(x)$ is the output from regression tree n. Considering its high predictive power, more and more authors have been using GBM to predict air quality [3], [4].

$$\hat{f}(x) = \sum_{n=1}^{N} \hat{f}^n(x). \tag{1}$$

All experiments are implemented with the Scikit-learn library (version 0.23.2). The tuning parameters are presented in Table 1. For this Scikit-learn version, we defined the learning rate, max_depth and random_state for all GBM models. The best results from GBM models are reported in the results section. The learning_rate was tuned to 0.05 to satisfy the convergence criterion quickly. The other parameters were used as the default values of the Scikit-learn library.

Table 1: Parameters for GBM in Scikit-learn library.

Parameters	Values
learning_rate	0.05
max_depth	5,10,15,30
random_state	1234

3.2 Ensemble deep learning

The EDL model is composed of given sub-DL models. Sub-model 1 and 2 are based on LSTM and BiRNN, respectively. While sub-model 3 is a combination of two LSTM layers, sub-model 4 is a combination of two BiRNN layers. Finally, sub-model 5 is a hybridization of BiRNN and LSTM layers.

In Fig. 2, the input data include the meteorological and temporal features. Next, 20 models (five sub-models × four "the number of nodes" in Table 2) were created for each pollutant. Afterwards, we select the top three best sub-models in the "meta learner" phase. The final output is computed by eqn (2), and this is the average output from these sub-models. Note that $\hat{f}^n(x)$ is the output from the three best sub-models. Therefore, N has a value equal to three.

$$\hat{f}(x) = \frac{1}{N}\sum_{n=1}^{N} \hat{f}^n(x). \tag{2}$$

The parameters for all the DL models are shown in Table 2. Tensorflow library, version 2.3.0, was used for all the DL sub-models. "The number of nodes" are changed to find the best model for each pollutant. "Epochs" parameter is a condition to stop the model. Specifically, the DL models will stop after 300 iterations if the model cannot obtain global optimization or improve the mean square error (MSE). By contrast, the model will cease the training phase by the "patience" parameter. Therefore, the training model finishes if the model accuracy cannot improve after 50 iterations (patience = 50). The "drop out" (20%) eliminates many connections between two layers to reduce the overfitting. The learning rate

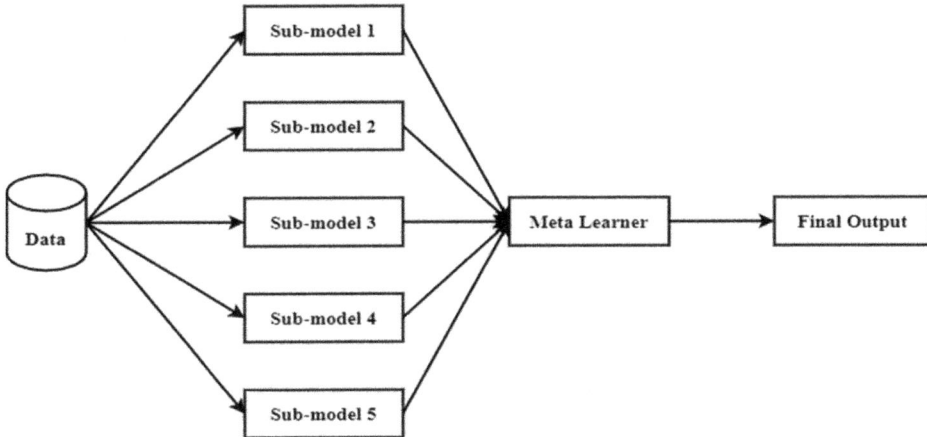

Figure 2: EDL model with its five sub-models.

Table 2: Parameters for all the DL models.

Parameters	Values
The number of nodes	16, 32, 64, 128
Patience	50
Drop out	0.2
Loss Function	mse
Learning_rate	0.05
Batch size	500
Epochs	300

is 0.05, which improves the convergent speed of the DL models. "Batch size" controls the number of training samples to update the weights in DL models.

3.2.1 Sub-model 1 and sub-model 2

In these models, the first layer is the input layer, which includes the meteorological and temporal features (Fig. 3). This layer transforms the raw format to the DL layer formats. In the sub-model 1, the next layer is the LSTM layer, which consists of many LSTM cells. The number of the LSTM cells are defined as "the number of nodes" (see Table 2). A "drop out" layer is introduced to reduce the overfitting of the model. Next, a dense layer with 20 nodes connects outputs from the "drop out" layer to the output layer. Since this is a binary classification, the Output Layer is a node with the sigmoid activation function to determine if the output belongs to peak ("1") or not peak ("0") levels of pollutant concentrations. The sub-model 2 (BiRNN) has a similar design to the LSTM sub-model. However, the LSTM cells and LSTM layer are replaced by the BiRNN cells and BiRNN layer.

3.2.2 Sub-model 3, sub-model 4 and sub-model 5

Fig. 4 represents the architecture of sub-model 3 (LSTM–LSTM) with its six layers. The input, "LSTM layer 1", "drop out layer", and the output layers are alike sub-model 1. We also tune the number of nodes for "LSTM layer 1". However, we added one more "LSTM

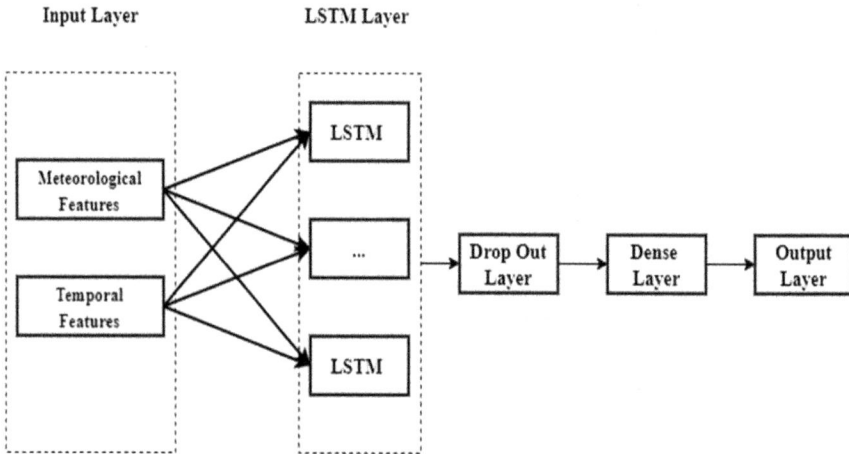

Figure 3: Sub-model 1 (LSTM).

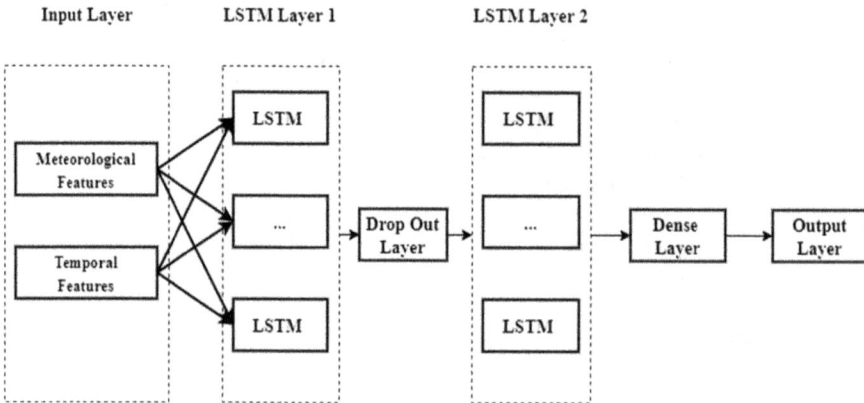

Figure 4: Sub-model 3 (LSTM–LSTM).

Layer 2" with 50 LSTM cells to obtain deeper information from the data. This layer is between the "drop out layer" and "dense layer". Additionally, dense layer has ten nodes and do not change when the parameters are tuned. In sub-model 4, the main difference is that the BiRNN cells in both BiRNN layers of sub-model 4 take the place of the LSTM cells in sub-model 3. Sub-model 5 is similar to sub-model 3, except for the fact that "LSTM layer 1" is BiRNN layer. The number of nodes in sub-model 4 and 5 is tuned similar to sub-model 3.

3.3 Metrics

Two metrics are used to compare the performance of each model: area under the ROC curve (AUC) and F1 score (or F-score). ROC is defined as a curve plotted from two parameters, which are the true positive rate (TPR) and false positive rate (FPR) resulting from the classification. The area under the ROC curve is the AUC value, and ranges from 0.5 to 1. *TPR* and *FPR* are computed by true positive (TP), false negative (RN), false positive (FP) and true negative (TN) as described in eqns (3) and (4). F1 score (eqn (7)) is based on the calculation of the *Precision* (eqn (5)) and the *Recall* (eqn (6)). The range values of the F1 score are from 0 to 1. Both metrics (AUC and F1) are calculated by the Scikit-learn library (version 0.23.2). The closer to 1, the better model is.

$$TPR = \frac{TP}{TP+FN},$$ (3)

$$FPR = \frac{FP}{FP+TN},$$ (4)

$$Precision = \frac{TP}{TP+FP},$$ (5)

$$Recall = \frac{TP}{TP+FN},$$ (6)

$$F1\ score = \frac{2\times precision \times recall}{precision+recall}.$$ (7)

3.4 SHapley additive exPlanations

Although DL has high accuracy, it is a kind of black-box which limits its interpretation and the identification of the weight of each feature in the prediction. Currently, two popular methods were introduced to disclose this opacity. The first one, SHAP [27], can provide the variable importance of input features in the DL models. These values are computed in terms of the game theory knowledge. The second, local interpretable model-agnostic explanations (LIME) uses coefficients among features to explain the model performance. However, it is tricky to choose the correct parameters from the LIME method, which can lead to miss significant variables [28]. Hence, we use SHAP for variable importance in this study.

SHAP values will allow us to get a more in-depth understanding of the contributions of the meteorological and temporal features to classify the level of pollution. In the EDL models, we applied SHAP for each sub-model. The SHAP values of EDL are the averages of SHAP values from the entire sub-models.

4 RESULTS AND DISCUSSION

4.1 Performance

Table 3 represents the average performance of all models. Overall, the EDL models outperform the other models. It is to note that EDL models are better than both GBM and sub-models in CO (F1 score = 0.8094; AUC = 0.8025), O_3 (F1 score = 0.8775; AUC = 0.8712), SO_2 (F1 score = 0.6969; AUC = 0.7126) and $PM_{2.5}$ (F1 score = 0.7079; AUC = 0.7195). However, even if the accuracy of the EDL model is always better than GBM, its AUC is slightly lower than sub-model 1 for NO_2. The reason is that the average of the five sub-models contributes to the final outputs of the EDL, as the top three sub-models are used for the calculation. Therefore, each selected sub-model contributes one third to the final classification.

It is to highlight the fact that the performance of EDL is higher than 0.7 for both metrics (only the F1 score of SO_2 is approximately 0.7 (0.6969) [29]. On the contrary, GBM scores are lower than 0.7 for SO_2 (F1 score = 0.7065; AUC = 0.6690) and $PM_{2.5}$ (F1 score = 0.6749; AUC = 0.6731). Hence, it can be concluded the EDL models are more reliable than both GBM and simple deep learning models to discriminate between daily peak or not daily peak levels of pollution concentration.

Table 3: Performance of GBM, EDL models with sub-models on testing set.

Pollutant	Metrics	GBM	Sub-model					EDL
			1	2	3	4	5	
NO$_2$	F1 score	0.7596	0.7736	0.7686	0.7611	0.7730	0.7641	0.7737
	AUC	0.7521	0.7638	0.7529	0.7485	0.7562	0.7512	0.7622
CO	F1 score	0.7769	0.7945	0.7894	0.7937	0.7979	0.7971	0.8094
	AUC	0.7726	0.7871	0.7858	0.7830	0.7855	0.7860	0.8025
O$_3$	F1 score	0.8729	0.8725	0.8562	0.8717	0.8714	0.8698	0.8775
	AUC	0.8671	0.8658	0.8501	0.8658	0.8644	0.8627	0.8712
PM$_{2.5}$	F1 score	0.7065	0.6911	0.6882	0.6888	0.7018	0.7008	0.7079
	AUC	0.6690	0.7082	0.7003	0.7038	0.7176	0.7179	0.7195
SO$_2$	F1 score	0.6749	0.6915	0.6622	0.6980	0.6842	0.6793	0.6969
	AUC	0.6731	0.6984	0.6934	0.7019	0.6967	0.6926	0.7126

To summarize, the best model performance in classifying daily peaks was found for O$_3$ and CO. The O$_3$ contaminant is one of the easiest pollutants to predict due to its high dependence on solar radiation activity, which always peaks at noon anywhere on the planet [30]. Some variations may be common due to increased cloudiness; however, those are also easy to predict using trends of relative humidity and atmospheric pressure. Similarly, CO may be affected by photochemical oxidation [31]. However, a week to months lifetime of CO (e.g., Holloway et al.), makes it an easier pollutant to predict, due to its longer persistence in the urban airshed [32].

On the contrary, NO$_2$, PM$_{2.5}$ and SO$_2$ are a bit less predictable, as they highly depend on anthropogenic emissions, but may also come from other sources [33]. Moreover, specific environmental conditions might help transport the emissions, or, on the other hand, might help the accumulation of air pollutants in an urban canopy. Variations in sources and the impact of meteorological parameters must be considered in addition to the complex trends of urban mobile circulation. While SO$_2$ can be emitted from high sulfur content fossil fuel burning, it can also be emitted by the active Andean volcanos. This latter factor is a completely unpredictable phenomenon given the set of parameters used in this study. This might help explain the poorest performance for this air pollutant.

4.2 Variable importance

SHAP is an advanced library for analyzing the inter-correlations among features of the GBM and DL models. Therefore, we used SHAP to identify the main factors that affect the concentration peaks of pollution. Fig. 5 represents the mean of the SHAP values for the five best EDL models (one for each pollutant). The higher the SHAP value is, the more important is the feature in the predicting output.

The results show that both the meteorological and temporal variables played a significant role in the best EDL models. Especially, "hours" is the most important variable in all five pollutant peaks' models, whereas the second and third crucial features are slightly different from one model to another. Since Bellisario is an urban area with heavy traffics, the pollution is highly affected by the time of the day. A previous study has identified two obvious pollution peaks at the rush hours (around 8 am and 6 pm) in the capital city of Ecuador [17].

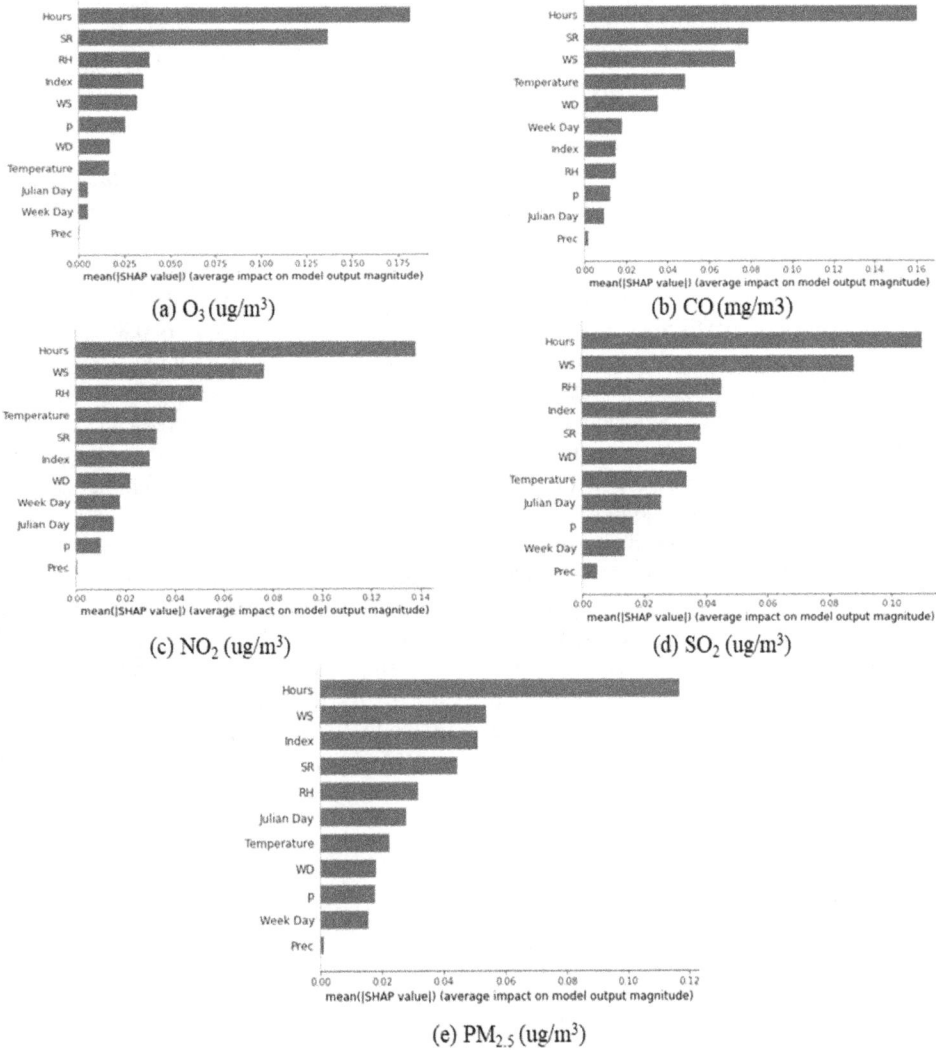

(a) O_3 (ug/m^3)

(b) CO (mg/m3)

(c) NO_2 (ug/m^3)

(d) SO_2 (ug/m^3)

(e) $PM_{2.5}$ (ug/m^3)

Figure 5:   Variable importance with mean (SHAP value) for the best models of five pollutants.

As discussed above, the most important predictor in all the models is the "hours". Daily trends of human activities carry the most important role in predicting air quality levels in this problematic high elevation city. While most of the pollutants might show the increased levels during the rush hours (low atmospheric mixing and increased traffic), O_3 shows a midday peak and thus also highly depends on SR (mean SHAP values ≈ 0.135) and next on RH (mean SHAP values ≈ 0.04) that might indicate cloud cover.

The second and third most important variables for CO are SR (mean SHAP values ≈ 0.08) and WS (mean SHAP values ≈ 0.07). As discussed above, CO is affected by photochemical oxidation, explaining the importance of SR. Now, WS is extremely important for all pollutants but O_3. WS is the above-mentioned ventilation parameter. An increase in wind speed can help ventilate the air pollutants, and transport them away from the emission sources, while the lack of it might create stagnant atmospheric conditions and generate an accumulation of anthropogenic urban emissions.

5 CONCLUSIONS

The prediction of atmospheric pollutant concentrations is always challenging to tackle, even if it is fundamental to anticipate on harmful effects of bad air quality on human health. It is even more difficult to forecast pollution peaks because they do not seem to follow any specific pattern. This study aims to determine the concentration peaks to get a better prediction of air pollution. Here, we propose an EDL approach to identify the daily peaks of concentration. The performance of EDL models ranges from 0.6969 to 0.8775, which demonstrates that the proposed method is promising for improving the prediction of pollution peaks. The highest accuracy is obtained for O_3, CO and NO_2 (above 0.75). As already reported in other studies, SO_2 and $PM_{2.5}$ are slightly more difficult to predict [34], [35]. According to the F1 and AUC metrics, we can conclude that the EDL outperforms both the traditional machine learning algorithm (GBM) and simple deep learning methods (LSTM and BiRNN) whatever the pollutants are considered.

The SHAP library allowed us to capture the interrelations among input features and pollutant concentrations in the EDL models. The time of the day (hours) – a marker of human activity tendencies – has been identified as the most significant feature in classifying the concentration peaks in all five contaminants. It can be explained by the fact that the capital city of Ecuador has two obvious daily peaks at the morning (8 am) and the evening rush hour (6 pm) [36]. Additionally, RH and WS are relevant factors in the NO_2 and SO_2 models. The second and third significant features depend on the nature of the pollutant. They are SR and RH for O_3, SR and WS for CO, WS and Index for $PM_{2.5}$. Further work will consist in using this classification to apply optimized models, specifically designed to predict high vs low concentrations.

ACKNOWLEDGEMENT

We thank the Secretariat of the Environment of the Municipality of the Metropolitan District of Quito for providing us with the air quality and meteorological data.

REFERENCES

[1]   Steinfeld, J.I., Atmospheric chemistry and physics: from air pollution to climate change. *Environment: Science and Policy for Sustainable Development*, **40**(7), p. 26, 1998. DOI: 10.1080/00139157.1999.10544295.

[2]   Barré, J., Petetin, H., Colette, A., Guevara, M., Peuch, V.H., Rouil, L., Engelen, R., Inness, A., Flemming, J., Pérez García-Pando, C. & Bowdalo, D., Estimating lockdown-induced European NO_2 changes using satellite and surface observations and air quality models. *Atmospheric Chemistry and Physics*, **21**(9), pp. 7373–7394, 2021.

[3]   Ceballos-Santos, S., González-Pardo, J., Carslaw, D.C., Santurtún, A., Santibáñez, M. & Fernández-Olmo, I., Meteorological normalisation using boosted regression trees to estimate the impact of COVID-19 restrictions on air quality levels. *International Journal of Environmental Research and Public Health*, **18**(24), p. 13347, 2021. DOI: 10.3390/ijerph182413347.

[4]   Rybarczyk, Y. & Zalakeviciute, R., Assessing the COVID-19 impact on air quality: A machine learning approach. *Geophysical Research Letters*, **48**(4), e2020GL091202, 2021. DOI: 10.1029/2020GL091202.

[5]   Grange, S.K., Carslaw, D.C., Lewis, A.C., Boleti, E. & Hueglin, C., Random forest meteorological normalisation models for Swiss PM$_{10}$ trend analysis. *Atmospheric Chemistry and Physics*, **18**(9), pp. 6223–6239, 2018.

[6]   Hochreiter, S. & Schmidhuber, J., Long short-term memory. *Neural Computation*, **9**(8), pp. 1735–1780, 1997.

[7]   Kuremoto, T., Kimura, S., Kobayashi, K. & Obayashi, M., Time series forecasting using a deep belief network with restricted Boltzmann machines. *Neurocomputing*, **137**, pp. 47–56, 2014.

[8]   Ong, B.T., Sugiura, K. & Zettsu, K., Dynamically pre-trained deep recurrent neural networks using environmental monitoring data for predicting PM$_{2.5}$. *Neural Computing and Applications*, **27**(6), pp. 1553–1566, 2016.

[9]   Navares, R. & Aznarte, J.L., Predicting air quality with deep learning LSTM: Towards comprehensive models. *Ecological Informatics*, **55**, 101019, 2020.

[10]   Krishan, M., Jha, S., Das, J., Singh, A., Goyal, M.K. & Sekar, C., Air quality modelling using long short-term memory (LSTM) over NCT-Delhi, India. *Air Quality, Atmosphere and Health*, **12**(8), pp. 899–908, 2019.

[11]   Schuster, M. & Paliwal, K.K., Bidirectional recurrent neural networks. *IEEE Transactions on Signal Processing*, **45**(11), pp. 2673–2681, 1997. DOI: 10.1109/78.650093.

[12]   Li, Q., Ness, P.M., Ragni, A. & Gales, M.J., Bi-directional lattice recurrent neural networks for confidence estimation. *IEEE International Conference on Acoustics, Speech and Signal Processing (ICASSP)*, pp. 6755–6759, 2019.

[13]   Tong, W., Li, L., Zhou, X., Hamilton, A. & Zhang, K., Deep learning PM$_{2.5}$ concentrations with bidirectional LSTM RNN. *Air Quality, Atmosphere and Health*, **12**(4), pp. 411–423, 2019. DOI: 10.1007/s11869-018-0647-4.

[14]   Zhang, Z., Zeng, Y. & Yan, K., A hybrid deep learning technology for PM$_{2.5}$ air quality forecasting. *Environmental Science and Pollution Research*, **29**, pp. 39409–39422, 2021. DOI: 10.1007/s11356-021-12657-8.

[15]   Catalano, M., Galatioto, F., Bell, M., Namdeo, A. & Bergantino, A.S., Improving the prediction of air pollution peak episodes generated by urban transport networks. *Environmental Science and Policy*, **60**, pp. 69–83, 2016.

[16]   Dutot, A.L., Rynkiewicz, J., Steiner, F.E. & Rude, J., A 24-h forecast of ozone peaks and exceedance levels using neural classifiers and weather predictions. *Environmental Modelling and Software*, **22**(9), pp. 1261–1269, 2007.

[17]   Zalakeviciute, R., Bastidas, M., Buenaño, A., Rybarczyk, Y., A traffic-based method to predict and map urban air quality. *Applied Sciences*, **10**(6), p. 2035, 2020.

[18]   WHO, *WHO Global Air Quality Guidelines*, 2021. https://apps.who.int/iris/bitstream/handle/10665/345329/9789240034228-eng.pdf?sequence=1&isAllowed=y. Accessed on: 6 Jan. 2022.

[19]   Zalakeviciute, R., López-Villada, J. & Rybarczyk, Y., Contrasted effects of relative humidity and precipitation on urban PM$_{2.5}$ pollution in high elevation urban areas. *Sustainability*, **10**(6), p. 2064, 2018. DOI: 10.3390/su10062064.

[20]   EMASEO, De Quito, Municipio del Distrito Metropolitano. Plan de desarrollo 2012–2022, 2011. *Quito: Municipio de Quito*.

[21]   INEC Q, Poblacion, superficie (km^2), densidad poblacional a nivel parroquial.

[22] Zalakeviciute, R., Rybarczyk, Y., López-Villada, J. & Suarez, M.V., Quantifying decade-long effects of fuel and traffic regulations on urban ambient $PM_{2.5}$ pollution in a mid-size South American city. *Atmospheric Pollution Research*, **9**(1), pp. 66–75, 2018. DOI: 10.1016/j.apr.2017.07.001.

[23] Henneman, L.R., Holmes, H.A., Mulholland, J.A. & Russell, A.G., Meteorological detrending of primary and secondary pollutant concentrations: Method application and evaluation using long-term (2000–2012) data in Atlanta. *Atmospheric Environment*, **119**, pp. 201–210, 2015.

[24] Enders, C.K., *Applied Missing Data Analysis*, Guilford Press, 2010.

[25] Lynch, S.M., *Introduction to Applied Bayesian Statistics and Estimation for Social Scientists*, Springer Science and Business Media, 2007.

[26] Friedman, J.H., Greedy function approximation: A gradient boosting machine. *Annals of Statistics*, **1**, pp. 1189–1232, 2001.

[27] Lundberg, S.M. & Lee, S.I., A unified approach to interpreting model predictions. *Advances in Neural Information Processing Systems,* **30**, 2017.

[28] Garreau, D. & Luxburg, U., Explaining the explainer: A first theoretical analysis of LIME. *International Conference on Artificial Intelligence and Statistics,* pp. 1287–1296, 2020.

[29] Luque, A., Carrasco, A., Martín, A. & de Las Heras, A., The impact of class imbalance in classification performance metrics based on the binary confusion matrix. *Pattern Recognition,* **91**, pp. 216–231, 2019.

[30] Juarez, E.K. & Petersen, M.R., A comparison of machine learning methods to forecast tropospheric ozone levels in Delhi. *Atmosphere*, **13**(1), p. 46, 2022. DOI: 10.3390/atmos13010046.

[31] Buchholz, R.R., Worden, H.M., Park, M., Francis, G., Deeter, M.N., Edwards, D.P., Emmons, L.K., Gaubert, B., Gille, J., Martínez-Alonso, S. & Tang, W., Air pollution trends measured from Terra: CO and AOD over industrial, fire-prone, and background regions. *Remote Sensing of Environment*, **256**, 112275, 2021. DOI: 10.1016/j.rse.2020.112275.

[32] Holloway, T., Levy, H. & Kasibhatla, P., Global distribution of carbon monoxide. *Journal of Geophysical Research: Atmospheres*, **105**(D10), pp. 12123–12147, 2020. DOI: 10.1029/1999JD901173.

[33] US EPA, Criteria air pollutants, 2022. https://www.epa.gov/criteria-air-pollutants. Accessed on: 13 Jan. 2022.

[34] Abdul-Wahab, S.A. & Al-Alawi, S.M., Prediction of sulfur dioxide (SO_2) concentration levels from the Mina Al-Fahal refinery in Oman using artificial neural networks. *American Journal of Environmental Sciences*, **4**(5), pp. 473–481, 2008. DOI: 10.3844/ajessp.2008.473.481.

[35] Li, J., Li, X., Wang, K. & Cui, G., Atmospheric $PM_{2.5}$ concentration prediction and noise estimation based on adaptive unscented Kalman filtering. *Measurement and Control.*, **54**(3–4), pp. 292–302, 2021. DOI: 10.1177%2F0020294021997491.

[36] Rybarczyk, Y. & Zalakeviciute, R., Regression models to predict air pollution from affordable data collections. *Machine Learning: Advanced Techniques and Emerging Applications*, pp. 15–48, 2018. DOI: 10.5772/intechopen.71848.

USE OF SATELLITE IMAGES TO ESTIMATE URBAN HEAT MAPS

ANA GABRIELA FERNÁNDEZ-GARZA[1], ERIC GIELEN[2] & JOSÉ-SERGIO PALENCIA-JIMÉNEZ[2]
[1]Planet and Sustainable Development, Universitat Politècnica de València, Spain
[2]Department of Urbanism, Universitat Politècnica de València, Spain

ABSTRACT
There is plenty of evidence about the change in the magnitude of climatic conditions in cities all over the world. It is an important problem especially since a large part of the world's population resides in cities. Comprehending the behaviour of Urban Heat Island (UHI) is necessary to plan cities that offset the climate change effects and offer greater security and quality of life to its inhabitants. In this way, planners need tools to model the UHI and research how cities interact with it. This paper aims to develop a methodological proposal to characterize UHIs in cities by using image data from the ASTER (Advanced Space Thermal Emissions and Radiometric Refection) satellite sensor. The case study is in the city of Valencia in Spain. It is based on data from three moments, for which day and night information are available for an identical ambit. The results provide six maps of the city of Valencia in a period of 4 years: a day and night one for 25 March 2017, 19 August 2018, and 7 May 2021. Hot and cool spots were detected in the city and the shape of the UHI was characterized. A few indicators were calculated to analyse them and study their evolution over time. Finally, we conclude that the use of image data from ASTER satellite offers a powerful infrastructure, easy and cheap to use, to model UHIs in cities. This methodology can be reproduced in many cities, as ASTER images are available for a large list of countries. This experience can be a useful tool in further study to obtain better knowledge about the relation between urban morphology and UHIs, so plan better cities.
Keywords: heat maps, urban form, Valencia, ASTER, Urban Heat Islands.

1 INTRODUCTION
Cities hold only 3% of the Earth's surface but are home to half of the world's population. Currently, 3.5 billion people live in cities and this number is still increasing, indeed, by 2030, it is estimated that 60% of the world's population would be living in cities [1]. The main effect of the climate crisis is the increase in the temperature of the Earth which is 1.1°C higher that at the end of the 19th century [2]. The effects of climate change, which has caused an increase of extreme events, make cities extremely vulnerable. The temperature rise is greater in the cities than in natural green spaces [3]. Therefore, the study of the relationship between temperature and the urban form of cities is especially relevant. First, it will be important to establish the causes to be able to provide the mitigation measures and second, it is important to achieve a configuration of cities where dwellings do not depend so much on air-conditioning.

Cities have their own microclimate, which is the result of changes in the environment through urban form, green spaces, infrastructures, and human activities. During the day, the objects accumulate heat on the contrary at night when they release [4]. This allows for establishing areas of higher temperature in cities compared to their immediate natural environment known as Urban Heat Islands (UHIs) [3].

However, estimating the urban temperature has the following limitations: low availability of stations, usually located outside the cities, so there is no data to establish the temperature inside the cities, and different behaviours of the materials and surfaces between day and night. One way to address these limitations is the use of satellite images, which can help to determine the land surface temperature (LST) of the entire city and the UHIs, based on the data of each pixel [5].

The UHI was estimated by the LST. The LST is the temperature emitted by the different surfaces [6], [7]. This temperature is different from the atmospheric temperature and has a direct relationship with the emissivity of land objects [8]. UHIs have been study with the use of satellite images to calculate the LST [9]–[13]. Those studies have been focussing in the comparisons between the values in-situ with the values obtained using the satellite images. In the case of Valencia City, a research with a pixel of 1,000 m (1 km) has established that there is not a significant result between the temperature and the typology of the buildings [13].

The objective of this work was to establish the relationship between LST and land use classification and test the use of the ASTER (Advanced Space Thermal Emissions and Radiometric Refection) satellite to do it. First, the satellite image is manipulated to produce an LST. Second, cold and hot spot areas are detected. Third, the relationship between LST and land use classification is studied with an analysis of variance (ANOVA) test. Finally, results are discuses by identifying the influence of land use on the UHIs of the cities.

2 METHODOLOGY

2.1 Scope

The city of Valencia is on the Mediterranean coast of Spain in the Comunidad Valenciana (Fig. 1). It had a population of 789,744 in January 2022. With 13,465 hectares, it is the centre of a metropolitan area that has more than one and a half million people. The temperature in Valencia varies from 6 to 30°C, and, in exceptional cases, it drops below 2°C or rises to more than 33°C. Although the major time of the year is dry, there are some differences in seasons: summers are hot, muggy, and mostly cloudless; winters are cold, windy, and partly cloudy.

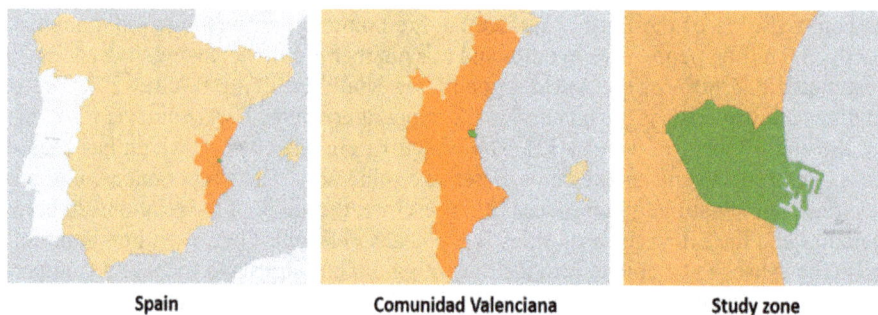

| Spain | Comunidad Valenciana | Study zone |

Figure 1: Study area.

The city of Valencia is a compact Mediterranean city with an important historic centre and extensive enlargement quarter city with high population density. The east limit is the Mediterranean Sea with important port infrastructure. In the north, west and south limits there is a protected orchard called L'Horta de Valencia (Miralles García, 2015). It is divided by an old riverbed, which currently corresponds to a large urban green area (Parque del Antiguo Cauce del Turia). For 50 years, the true river has been displaced in the southern limit of the city (Fig. 2). According to land use classification [14], low-density residential, bare soil, and supply infrastructure take little surface in the city.

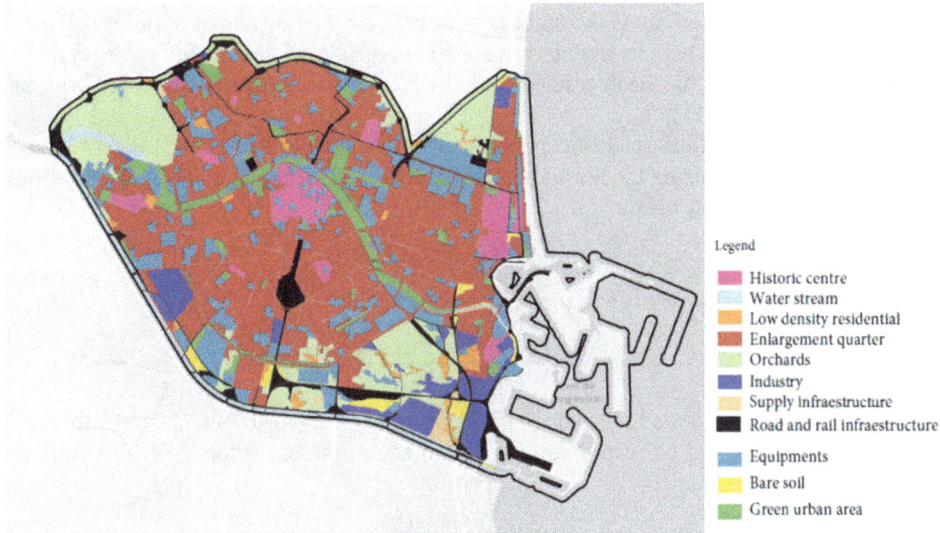

Figure 2: Land use distribution in 2014 [14].

2.2 Methods

First, images without clouds in the study area from the ASTER satellite were selected ASTER is the only high spatial resolution instrument on the Terra platform, designed, calibrated, and validated by a joint United States/Japan team The ASTER obtains high-resolution (15–90 m^2 per pixel) images of the Earth in 14 different wavelengths of the electromagnetic spectrum, ranging from visible to thermal infrared light [5].

In scientific literature, ASTER data has been used to create detailed maps of LST, emissivity, reflectance, and elevation. One study compares the results of each band in the island of Tenerife, Spain, to determine which of the bands had the greatest agreement. The study concludes that bands 13 and 14 have a difference of squares less than 0.002, while bands 10 and 11 have one of 0.015, being better than bands 13 and 14. In addition, the value cannot be obtained for band 12 because band 12 have problems [5].

From the 14 bands of ASTER image, only three were used: band 2 and 3 from the visible and near-infrared (VNIR) with 15 m resolution, and band 13 from the temperature subsystem (thermal infrared) with 90 m resolution. The band 13 (10.25–10.95 μm) was chosen, since bands 10 and 14 are close to the atmospheric window having more affection from the atmospheric effects. Additionally, working with band 12 was discarded too, since its sensitivity has decreased over time [5].

The LST was calculated following eqn (1):

$$LST = \frac{T_b}{\left[1+\left(\frac{\lambda \cdot T_b}{C_2}\right)\cdot Ln_{LSE}\right]},\tag{1}$$

where LST = land surface temperature; T_b = apparent brightness temperature; $C_2 = h \cdot \frac{c}{s} = 1.4388 \cdot 10^{-2} \, m \, K$; h = Planck's constant: $6.626 \cdot 10^{-34}$ Js; c = speed of light: $2.998 \cdot 10^8$ m/s; s = Boltzman's constant = $1.38 \cdot 10^{-23}$ J/K; LSE = land surface emissivity or "ε"; Λ = central wavelength of the thermal band emitted for band 13 is 10.6 m [4], [6], [7].

The land surface emissivity is the amount of energy emitted by a material of an object at the same temperature [12]. It is determined by atmospheric and material factors. The estimation of emissivity has been studied with different algorithms and compared with on-site measurements [9]–[11].

Based on these previous scientific papers and the available data, it was determined to calculate the emissivity using the normalized difference vegetation index (NDVI) thresholds method algorithm [12], as in eqn (2):

$$\varepsilon_{13} = 0.968 + 0.0022 \cdot Pv. \tag{2}$$

The vegetation fraction Pv is obtained [7], as in eqn (3):

$$Pv = \left[\frac{NDVI-NDVI_{min}}{NDVI_{max}-NDVI_{min}}\right]^2. \tag{3}$$

NDVI values are calculated as in eqn (4) [12], using band 3N for NIR and band 2 for RED, and where NDVImax and NDVImin are maximum and minimum values of pixels from the image.

$$NDVI = \frac{NIR-RED}{NIR+RED}. \tag{4}$$

This methodology was applied to a group of three couples of day/night images. As the night images do not have the NIR and RED bands, the land surface emissivity determined for the day was used for the night [8], [15].

Calculations were solved in QGIS [16], with the semi-automatic classification plugin (SCP). Finally, the raster image was converted to points to help next data manipulation and analysis.

After calculating the LST, several statistics techniques were used to study both the spatial autocorrelation of temperature values and analyses of variance of Land Use temperature. First, a Moran's I test was carried out to define if the data is clustered, and therefore, there is not a random pattern. The test gives a p-value as the probability that the spatial pattern was generated by a random process: if the p-value is low, it is unlikely that the spatial pattern is the result of random processes, so the null hypothesis can be rejected. The Moran's I Index give values between –1 and 1, when the value is positive it means that high values cluster near other high values and the same for the low values. In the case when the values are negative, it occurs when the high and low values are close together. The value is close to zero when the positive values balance the negative values [17], [18]. Second, having determined that the spatial pattern is not random, a hot spot analysis with the Getis-ord Gi statistic [17] were used, giving a z-scores (GiZScore) and p-values, which tell you where features with either high or low values cluster spatially. GiZScore values were classified into seven groups created by quantile (equal count) classification and then the hot spots (seventh group) were analyzed with meteorological data, to establish patterns and singularities. Finally, LST and land use classification comparison intersecting both in QGIS and with an ANOVA test and a post hoc tests to evaluate the means equality of the land use temperature. Land use classification data is extracted in 2014 from SIOSE database [14]. ANOVA one-direction analysis allows verifying the equality of the means. Then, with this first ANOVA test done, a second post-test called post hoc test is applied to check which land use has a significantly different means from others. In the post hoc test, each of land uses is compared in pairs with another to determine which set of land uses has equal or different means. When the p-value is lower than 0.05, the null hypothesis is rejected, so it can be established that the means of both land's uses are significantly different [19]–[21].

3 RESULTS

Three couples of ASTER images with day and night data were studied: March 2017, August/November 2018 and May 2021 (Table 1). As it seems important to have meteorological data to interpret LST, Table 2 gives air temperature and atmospheric pressure, speed, and direction of the wind for each of the images.

Table 1: ASTER images selected.

Day		Night	
Date – hour	Aster image	Date – hour	Aster image
25 March 2017 – 10:55:00	RT_AST_L1T_ 00303252017105447_ 20170326091144_12025	16 March 2017 – 22:05:31	RT_AST_L1T_ 00303162017220536_ 20170317085447_24894
19 August 2018 – 10:56:10	RT_AST_L1T_ 00308192018105610_ 20180820133206_5393	23 November 2018 – 21:59:45	RT_AST_L1T_ 00311232018215945_ 20181127171037_27959
7 May 2021 – 10:52:49	RT_AST_L1T_ 00305072021105254_ 20210508111845_18908	7 May 2021 – 21:57:25	RT_AST_L1T_ 00305072021215732_ 20210508092259_8521

Table 2: Meteorological data for Valencia city [22].

	Date	Hour	Air temperature (°C)	Pressure atmospheric (mnbar)	Wind (km/h and direction)
a	25 March 2017	10:30	13	1.016	18.4 O
b	16 March 2017	22:00	11	1.023	5.40 NO
c	19 August 2018	10:30	25	1.021	3.6 NE
d	23 November 2018	21:30	10	1.016	13 SO
e	7 May 2021	10:30	20	1.016	5.40 E
f	7 May 2021	21:30	21	1.017	13 NE

LST maps of Valencia City for each date were calculated (Fig. 3). The blue colour is for low temperature and the red colour for the high one. An important effect of orchards and urban green areas (Parque del Antiguo Cauce del Turia) can be seen with lower value, even more, visible at night. At night, the effect of the density of blocks in the residential city and concrete infrastructure is visible too. At the east, the influence of the Valencia port is very high with higher temperatures. It is also necessary to highlight in the south, the hotspot corresponding to the new river, with a particular configuration in the case of Valencia City: a fluvial channel without water for most of the year plus a three-lane highway on each side.

Spatial autocorrelation test (global Moran's I) gives p-value equal to zero for all the images (Table 3): it is very unlikely (small probability) that the observed spatial pattern is the result of random processes. The p-value is statistically significant, and the z-score is positive. The spatial distribution of high values and/or low values in the image's dataset is more spatially clustered than would be expected if underlying spatial processes were random.

Although spatial autocorrelation exists in all the images, during the day, the heat islands are more difficult to observe: hot and cold spots are more dispersed in specific locations. Lower Moran's Index and z-score for day images in March 2017 and August 2018 seem to confirm it. Hot spots can be seen associated with infrastructures, industrial areas, and

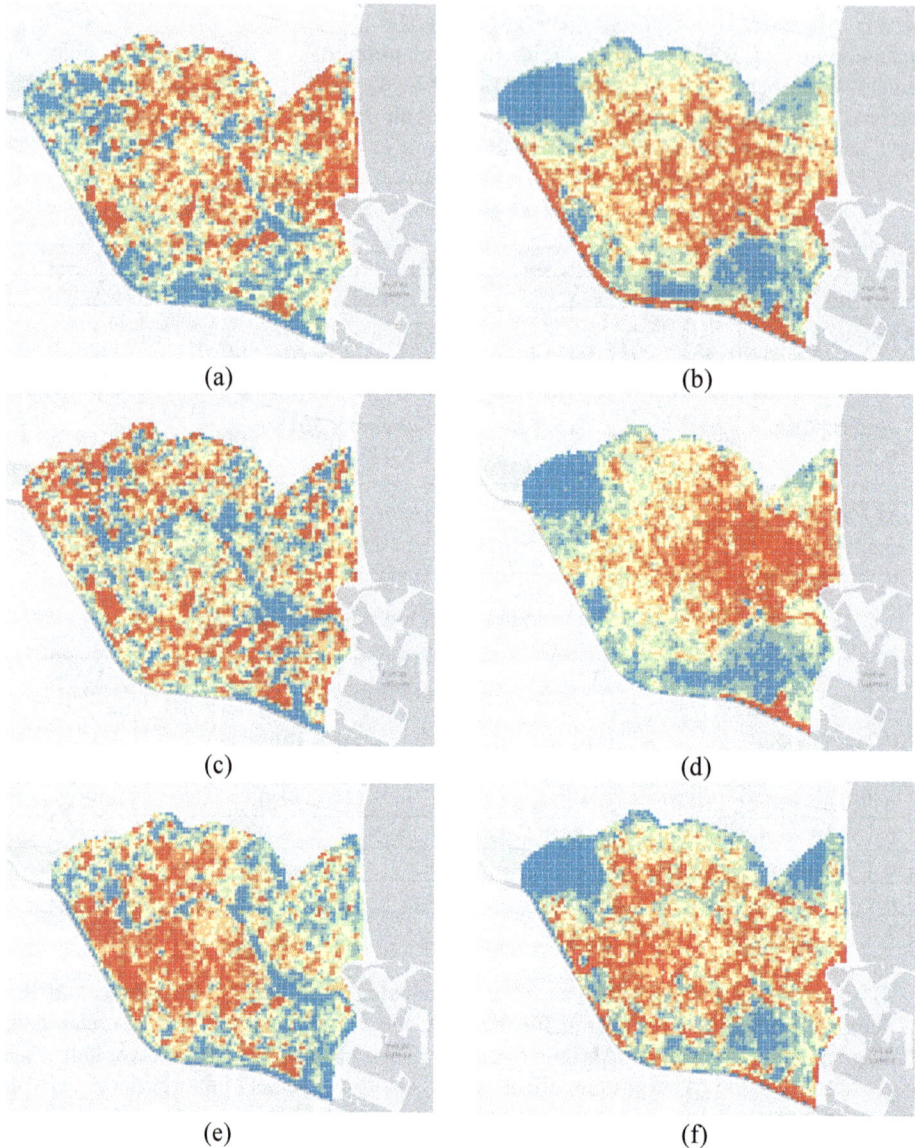

Figure 3:   Land surface temperature. (a) 25/ March 2017; (b) 16 March 2017; (c) 19 August
2018; (d) 23 November 2018; (e) 7 May 2021; and (f) 7 May 2021.

facilities distributed in the town. There appear associated for example with some roofs of
industrial buildings or facilities, as well as with the artificial grass of sports centres and soccer
fields. Orchards and open spaces all around the city present a more heterogeneous distribution
of temperature too: in the case of orchards, it is made up of a mosaic of crops with very
different vegetative states giving very different responses to the sun. On the contrary, at night,
heat islands are easier to see as it corresponds more obviously to the urban form of Valencia
City.

Fig. 4 and Table 4 present results of temperature according to land use classification with a boxplot diagram where differences can be seen according to land use classification.

Table 3: Results for global Moran's I test for each land surface temperature.

Date – Hour	Moran's Index	Expected Index	Variance	z-score	p-value
25 March 2017 – 10:55:00	0.056554	–0.000149	0	204.797358	0
16 March 2017 – 22:05:31	0.108456	–0.000149	0	391.721004	0
19 August 2018 – 10:56:10	0.024697	–0.000149	0	90.126541	0
23 November 2018 – 21:59:45	0.153994	–0.000149	0	556.080158	0
7 May 2021 – 10:52:49	0.08186	–0.000149	0	296.301856	0
7 May 2021 – 21:57:25	0.080565	–0.000149	0	292.207661	0

Then to confirm if the differences between group means are statistically significant, an ANOVA test is done. Results in Table 5 confirm with $p < 0.001$ for all the images, that differences among land use categories are significant.

Table 6 details the post hoc test for the night image of May 2021. This test allows us to confirm category pairs with different means:

- Historic centre registers significant difference in mean temperature than all categories, except with the supply infrastructure. At night, the historic centre temperature is 2.2°C bigger than orchards, 0.3°C versus Green urban area, 2.5°C versus water stream, 0.7°C then low-density residential form, 0.5°C versus industry, and 0.7°C then road and rail infrastructure; while it is 0.2°C lower than enlargement of the city.
- Enlargement has a significant positive difference from other urban uses, except for the supply infrastructure. Differences vary in an interval from +0.2°C with the historic centre to 2.4 and 2.7°C for orchards and water stream, respectively: historic centre (0.2°C), equipment (0.4°C), green urban area (0.5°C), industry (0.8°C), road and rail infrastructure (0.9°C), orchards (2.4°C), and water stream (2.7°C). The enlargement city of Valencia is a higher-density urban form constructed in the last century.
- Green urban area has a significant difference from other uses in the city, except with supply infrastructure, equipment, and bare soil. It is lower than in the historic centre (–0.3°C) and enlargement city (–0.5°C). On the contrary, the temperature of green urban areas is bigger than water stream (2.3°C), orchards (1.9°C), road and rail infrastructure (0.4°C), low-density residential form (0.4°C), and industry (0.3°C)
- Orchards use has significant differences except with water stream. Both uses are spatially located outside the urban area; they define the limits of the city, a cold spot all-around Valencia city that contributes to cold the urban area. Orchard's temperature is lower than urban types, among –1.5°C to –2.4°C: enlargement city (–2.4°C), supply infrastructure (–2.3°C), historic centre (–2.2°C), equipment (–2°C), green urban area (–1.9°C), industry (–1.6°C), road and rail infrastructure (–1.5°C), and low residential density form (–1.5°C).

4 CONCLUSIONS

ASTER images seem to be useful for obtaining urban heat maps due to their resolution (90 m) and their wide temporal and spatial availability. With a methodology as proposed in the study, it is possible monitoring the evolution of the heat island, so that it can be learnt

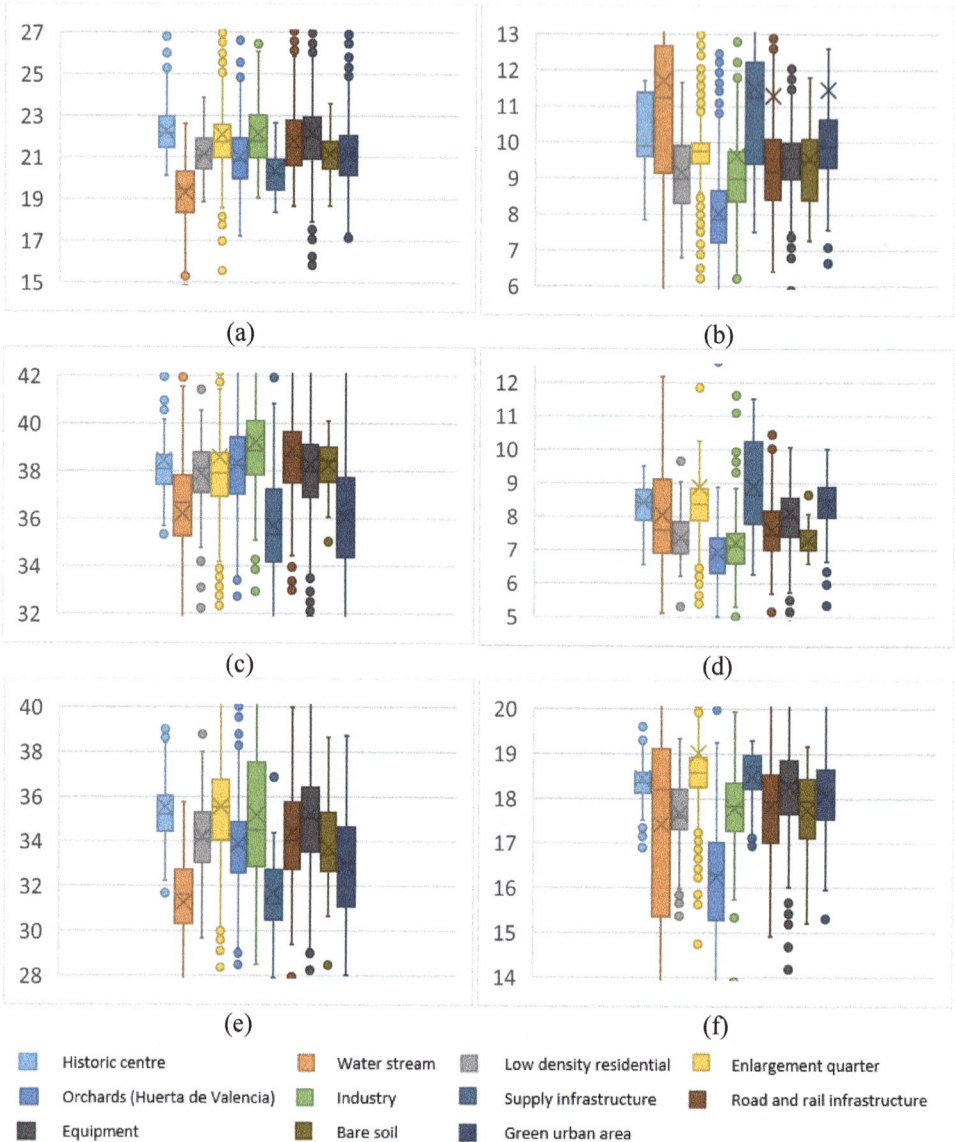

Figure 4: Land surface temperature according to land use classification. (a) 25 March 2017; (b) 16 March 2017; (c) 19 August 2018; (d) 23 November 2018; (e) 7 May 2021; and (f) 7 May 2021.

more about the interaction between urban form and climate change effects in the city. Comparing day and night, the study of urban heat gives better results at night with more accurate identification and delimitation of the heat island. During the day, there are many outsiders, with high temperatures due to different responses of specific materials to solar radiation, complicating the interpretation of the map.

Table 4: Land surface temperature according to land use classification. (a) Temperature media; and (b) Difference with median temperature.

Land use classification	25 March 2017 – 10:55:00		16 March 2017 – 22:05:31		19 August 2018 – 10:56:10		23 November 2018 – 21:59:45		7 May 2021 – 10:52:49		7 May 2021 – 21:57:25	
	a	b	a	b	a	b	a	b	a	b	a	b
Historic centre	22.3	1.1	9.9	0.3	38.5	0.2	8.4	0.4	35.6	1.3	18.5	0.6
Water stream	19.3	–1.9	11.7	2.1	36.2	–2.1	8.1	0	31.3	–3	17.4	–0.4
Low density residential	21.1	0	9	–0.6	37.9	–0.4	7.4	–0.7	34.2	0	17.7	–0.2
Enlargement quarter	22.1	0.9	9.7	0.1	38.6	0.3	8.9	0.8	35.6	1.4	19	1.2
Orchards (Huerta de Valencia)	21	–0.2	7.8	–1.8	38.3	0	6.9	–1.1	33.9	–0.3	16.2	–1.6
Industry	22.1	1	9	–0.6	39.2	0.9	7.2	–0.8	35.2	1	17.8	0
Supply infrastructure	20.3	–0.8	11.3	1.7	35.7	–2.6	8.9	0.9	31.7	–2.5	18.5	0.7
Road and rail infrastructure	21.8	0.6	9.2	–0.4	38.9	0.6	7.6	–0.4	34.4	0.2	17.8	0
Equipment	22.2	1	9.6	0	38.3	0	8.1	0.1	35.2	1	18.3	0.4
Bare soil	21.2	0	9.6	0	38.3	0	7.3	–0.8	33.7	–0.5	17.7	–0.1
Green urban area	21.1	0	9.9	0.3	36.1	–2.2	8.4	0.3	32.9	–1.3	18.1	0.2

Table 5: Results of one-way ANOVA and homogeneity of variances test (Levene's).

Date	One-way ANOVA						Homogeneity of Variances Test (Levene's)			
	Day/Night	Prueba	F	df1	df2	p	F	df1	df2	p
25 March 2017	Day	Welch's	84.6	10	587	<.001	18.5	10	6694	<.001
		Fisher's	79.7	10	6694	<.001				
16 March 2017	Night	Welch's	366	10	576	<.001	230	10	6694	<.001
		Fisher's	323	10	6694	<.001				
19 August 2018	Day	Welch's	52.6	10	582	<.001	212	10	6694	<.001
		Fisher's	56.8	10	6694	<.001				
23 November 2018	Night	Welch's	363	10	582	<.001	117	10	6694	<.001
		Fisher's	278	10	6694	<.001				
7 May 2021	Day	Welch's	130	10	585	<.001	20.3	10	6694	<.001
		Fisher's	110	10	6694	<.001				
7 May 2021	Night	Welch's	285	10	579	<.001	216	10	6694	<.001
		Fisher's	297	10	6694	<.001				

Table 6: Results of post hoc tests for 07-05-2021 night image. (a) Mean difference; and (b) p-value.

Land use classification		1	2	3	4	5	6	7	8	9	10	11
Historic centre (1)	a		2.53	0.67	-0.21	2.16	0.55	-0.15	0.67	0.14	0.59	0.25
	b		<.001	<.001	<.001	<.001	<.001	0.937	<.001	0.005	<.001	<.001
Water stream (2)	a			-1.85	-2.73	-0.36	-1.97	-2.68	-1.85	-2.38	-1.93	-2.26
	b			<.001	<.001	<.001	<.001	<.001	<.001	<.001	<.001	<.001
Low density residential (3)	a				-0.88	1.49	-0.11	-0.82	0	-0.52	-0.07	-0.41
	b				<.001	<.001	<.001	<.001	1	<.001	1	<.001
Enlargement quarter (4)	a					2.37	0.76	0.05	0.881	0.352	0.80	0.46
	b					<.001	<.001	1	<.001	<.001	<.001	<.001
Orchards (Huerta de Valencia) (5)	a						-1.61	-2.31	-1.49	-2.02	-1.57	-1.90
	b						<.001	<.001	<.001	<.001	<.001	<.001
Industry (6)	a							-0.70	0.11	-0.41	0.04	-0.29
	b							<.001	0.98	<.001	1	<.001
Supply infrastructure (7)	a								0.82	0.29	0.74	0.41
	b								<.001	0.25	<.001	0.031
Road and rail infrastructure (8)	a									-0.52	-0.07	-0.41
	b									<.001	1	0.001
Equipment (9)	a										0.45	0.11
	b										0.01	0.46
Bare soil (10)	a											-0.33
	b											0.19
Green urban area (11)	a											
	b											

The urban heat maps obtained, and the spatial autocorrelation test done, show the existence of a heat island corresponding to Valencia City. This heat island is marked even more by the presence around the city of the Huerta de Valencia and the riverbed to the south.

The analysis of land use classification and its temperature confirms something similar. There are significant differences between land-use types, especially between artificial land use and open spaces such as orchards or urban green areas. There are also differences among urban types, although, in this case, surely the results could have been affected by the land-use dataset used that did not distinguish enough the urban forms. The difference between residential, industry, and infrastructure are more contrasted during the day, with important outsiders, while these differences are less at the night. Some use such as block roofs, pavements or artificial grasses shows that a lot of heat accumulates there during the day. Anyway, uses with residential blocks, industry, facilities, and concrete infrastructures contribute to rising temperature, while orchards and urban green areas record lower temperatures and bring down the immediate environment.

Urban heat maps have an easier interpretation at night and seem to work better.

ACKNOWLEDGEMENT
This research was funded by Planet and Sustainable Development Chair (Cátedra Planeta y Desarrollo Sostenible) of the Universitat Politècnica de València.

REFERENCES
[1]   United Nations, https://www.un.org/sustainabledevelopment/es/cities/. Accessed on: 13 Jun. 2022.
[2]   United Nations, https://www.un.org/es/climatechange/what-is-climate-change#:~:text =Las%20concentraciones%20de%20gases%20de,fue%20la%20m%C3%A1s%20c%C3%A1lida%20registrada. Accessed on: 13 Jun. 2022.
[3]   Weber, N., Haase, D. & Franck, U., Zooming into temperature conditions in the city of Leipzig: How do urban built and green structures influence earth surface temperatures in the city? *Science of the Total Environment*, **496**, pp. 289–298, 2014.
[4]   Weng, Q., Rajasekar, U. & Hu, X., Modeling urban heat island and their relationship with impervious surface and vegetation abundance by using ASTER imagens. *IEEE Transactions on Geoscience and Remote Sensing*, **49**(10), pp. 4080–4089, 2011.
[5]   Nichol, J., Fung, W., Lam, K.-S. & Wong, M., Urban heat island diagnosis using ASTER satellite images and 'in situ' air temperature. *Atmospheric Research*, **II**(94), pp. 276–284, 2009. DOI: 10.1016/j.atmosres.2009.06.011.
[6]   Feizizadeh, B. & Blaschke, T., Examining urban heat island relations to land use and air pollution: Multiple endmember spectral mixture analysis for thermal remote sensing. *IEEE Journal of Selected Topics in Applied Earth Observations and Remote Sensing*, **6**(3), pp. 1749–1756, 2013.
[7]   Bravo, N., *Teledetección Espacial LANDSAT; SENTINEL 2; ASTER L1T and MODIS*, 1st ed., Geomática Ambiental S.R.L.: Huánuco, Perú, 2017.
[8]   Jin, M. & Llang, S., An improved land surface emissivity parameter for land surface models using global remote sensing obervations. *Journal of Climate*, **19**, pp. 2867–2881, 2006.
[9]   Mira, M. & Coll, C., Evaluation of surface temperature and emissivity derived from ASTER data: A case study using ground-based measurements at a volcanic site. *American Meteorological Society*, **27**(10), pp. 1677–1688, 2010.

[10] Sobrino, J., Jiménez-Muñoz, J. & Paolini, L., Land surface temperatue retrieval from LANDSAT TM 5. *Remote Sensing of Enviromente*, **90**, pp. 434–440, 2004. DOI: 10.1016/j.rsc.2004.02.003.

[11] Oltra-Carrió, R., Sobrino, J., Franch, B. & Nerry, F., Land surface emissivity retrieval from airborne sensor over urban areas. *Remote Sensing of Environment*, **123**, pp. 298–305, 2012. DOI: 10.1016/j.rse.2012.03.007.

[12] Ndossi, M. & Avdan, U., Inversion of land surface temperature (LST) using terra ASTER data: A comparison of three algorithms. *Remote Sensing*, **8**(993), pp. 1–19, 2016.

[13] Cuesta Navarro, J., Caracterización de la Isla de Calor Urbana (ICU) mediante el uso de imágenes obtenidas por satélite, procesandas mediante software de código abierto QGis. Aplicación al caso de Valencia, Valencia, 2020.

[14] Instituto Geográfico Nacional, Sistema de Ocupación del Suelo de España (Cartografía Digital), 1:25.000, Madrid, 2014.

[15] Alipour, T., Sarajian, M. & Esmaeily, A., Land surface temperature estimation from thermal band of Landsat sensor, case study: Alashtar city. *The International Archives of the Phorogrammetry*, **XXXVIII**, 2004.

[16] QGIS Development Team, QGIS Geographic Information System, 2022. https://qgis.org.

[17] Gestis, A. & Ord, J., *Geographical Analysis*, vol. 24, 1992.

[18] Mitchell, A., *La Guìa de Esri para el análisis SIG*, vol. II, Esri Press, 2005.

[19] Fox, J. & Weisberg, S., Companion to applied regression. R package. 2020. https://cran.r-project.org/package=car.

[20] R Core Team, R: A language and environment for statistical computing. Computer software, Version 4.0, 2021. https://cran.r-project.org.

[21] The Jamovi Project, jamovi. Computer software, Version 2.2. 2021. https://www.jamovi.org.

[22] Weather Spark, Weather Spark, 2018. https://es.weatherspark.com/y/42614/Clima-promedio-en-Valencia-Espa%C3%B1a-durante-todo-el-a%C3%B1o. Accessed on: 17 Jun. 2022.

SECTION 2
GLOBAL, REGIONAL
AND LOCAL STUDIES

IMPORTANCE OF COMPREHENSIVE HEALTH RISK ASSESSMENT PROCEDURES FOR MODERN WASTE-TO-ENERGY FACILITIES IN COMPLEX GEOGRAPHICAL CONTEXTS ORIENTED TO CIRCULAR ECONOMY

ELENA CRISTINA RADA[1], MARCO TUBINO[2], MARCO SCHIAVON[3] & LUCA ADAMI[2]
[1]Department of Theoretical and Applied Sciences, University of Insubria, Italy
[2]Department of Civil, Environmental and Mechanical Engineering, University of Trento, Italy
[3]Department of Agronomy, Food, Natural Resources, Animals and Environment, University of Padova, Italy

ABSTRACT

Although circular economy (CE) principles set material circularity, resource efficiency and waste recycling as priority targets to guarantee the sustainable development of future generations, the thermo-chemical valorisation of municipal solid waste (MSW) still plays a fundamental role in the transition towards the final CE targets. As a matter of fact, the waste-to-energy (WtE) sector allows recovering energy from waste, reducing the pressure on MSW landfills and their related potential environmental impacts; however, recovering material for further uses is not excluded in WtE options. Significant improvements have been achieved in the air pollution control of exhaust gases from direct and indirect MSW combustion during the last decades. The efforts focussed on reducing dioxin emissions especially, and this has let other substances emerge as priority pollutants (e.g., heavy metals). In addition, the location of WtE facilities in certain geographical contexts is still potentially critical from the point of view of human exposure and the related health risk; moreover, the public acceptance of WtE plants is still limited, in spite of their recent role in fighting SARS-CoV-2 risks from waste management. The purpose of the present paper is to underline the importance of implementing correct and complete health risk assessment procedures tailored to the exposed population living in the area of influence of a WtE plant. The paper will present two case studies regarding the projects of two WtE plants in a mountainous region, highlighting the critical issues that arose during the environmental impact assessment procedures. The paper will finally suggest possible options to improve the health risk assessment procedure and alternative measures to reduce the expected impacts of the WtE sector on the environment and human exposure.
Keywords: air pollution, cancer risk, circular economy, emissions, environmental impact assessment, heavy metals, human exposure, incineration, gasification.

1 INTRODUCTION

The atmospheric dispersion of air pollutants emitted from ground-level human activities is particularly limited in mountainous regions. This is due to the complex morphology of mountainous areas, where valley winds (especially down-slope) induce strong local circulations that hinder the vertical air mixing normally occurring in flat areas [1]–[3]. In mountainous regions, this translates into enhanced atmospheric stability conditions, which develop especially at night and during winter, conditions that tend to trap air pollutants near the ground [4], [5]. The main sources of air pollutants are human activities (e.g., road traffic, industries, domestic heating), which are usually located at the valley bottom [6]–[11]. Besides being the most critical part of the year in terms of atmospheric dispersion, winter is the season with the highest emissions of air pollutants, due to the additional contribution of domestic heating. The use of wood-based biomass for domestic heating, especially, is widespread in mountainous areas, where wood biomass is relatively abundant in continental climates, and is a known source of particulate matter, nitrogen oxides, polycyclic aromatic hydrocarbons, dioxin and polychlorinated biphenyls [6], [9], [12]–[14]. Thus, air pollutants

WIT Transactions on Ecology and the Environment, Vol 259, © 2022 WIT Press
www.witpress.com, ISSN 1743-3541 (on-line)
doi:10.2495/AWP220051

are released in a critical atmospheric layer where the exchange of air with the upper layers is limited.

The presence of population, normally settled at the valley bottom, exacerbates this problem, since the stagnation of air pollutants near the ground may become a threat for human health. In large mountainous contexts, like the Alpine region, the presence of urban areas with relatively high population density requires appropriate waste management plans to deal with the municipal solid waste (MSW) produced. Under the vision of the European Waste Framework Directive (2008/98/EC) [15], priority should be given to politics favouring waste prevention, preparing for re-use and waste recycling in concordance with the circular economy (CE) principles [16]–[19]. Under this scheme, waste-to-energy (WtE) processes are considered as the last viable option before landfilling. In this context, WtE processes are intended as thermo-chemical processes applied to non-biodegradable waste (e.g., refuse-derived fuel, residual fraction of MSW or waste assimilated to MSW) as complementary to biorefineries. Along the pathway that leads toward the full application of CE principles, the WtE sector plays a key role in waste management, as it allows: (1) reducing the pressure on MSW landfills, (2) reducing the environmental impacts of landfills, (3) recycling material from bottom ash and (4) replacing fuels for energy production [20].

From the point of view of environmental impacts, the WtE sector has been significantly improved in the last decades. The best results have been achieved in the emission control of polychlorinated dibenzo-p-dioxin and dibenzo-furans (PCDD/Fs) and polychlorinated biphenyls (PCBs), as confirmed also by long-term exposure assessment studies in populations living near waste combustion plants [21], [22]. Following such improvements, other pollutants, which were previously considered of secondary importance, should now receive priority in terms of emission control, environmental monitoring and health risk assessment. The contribution of heavy metals and, especially, hexavalent chromium (Cr VI) to the total cancer risk from WtE plants was recently pointed out [23]. In addition, despite the advances in terms of PCDD/F and PCB emission control, the social acceptance of WtE plants is still limited and influenced by the so-called "not in my backyard" (NIMBY) syndrome [24].

The impacts on human health of the pollutants emitted from civil and industrial activities are estimated through the environmental impact assessment (EIA) process, which was developed and implemented in the United States in 1970 with the National Environmental Policy Act [25]. The assessment of the risk for health, as part of the EIA process, requires caution, because the results may be influenced by the choices and hypotheses made to simplify the evaluation, such as the selection of the target area, the emission sources to consider, the exposure routes, the morphologic and meteorological data, the resolution of the computational domain for meteorological and dispersion simulations [26]. Thus, the development of correct health risk assessment procedures becomes crucial to avoid misleading results, potential underestimations of the impacts of human activities on health and unnecessary precautionary approaches that may lead to socio-economic damages to local communities. Correct EIA procedures have also the power to increase the level of social acceptance of WtE, as recently observed in China [27]. Accurate health risk assessment procedures are even more important in geographical contexts that may amplify the impacts expected by one source, like valleys.

The present paper aims at highlighting the importance of the health risk assessment process in the evaluation of the potential risks involved when authorising WtE plants based on direct or indirect waste combustion. To facilitate the understanding of the critical issues that may emerge when carrying out an EIA process, two case studies will be presented and discussed. Both case studies refer to waste combustion plants that were proposed for authorisation in an Alpine valley, but that were not realised in the area. The paper will propose possible measures to reduce the environmental impacts from the WtE sector and, meanwhile, improve the risk assessment procedure and the environmental legislation in order to properly account for potential underrated risks for health. The role that WtE are playing during the SARS-CoV-2 does not affect the methodology presented in the following sections.

2 CASE STUDIES

The selected case studies refer to two waste combustion plants that were proposed in different locations of the same valley (Adige Valley), located in northern Italy in the Italian Alps (Fig. 1).

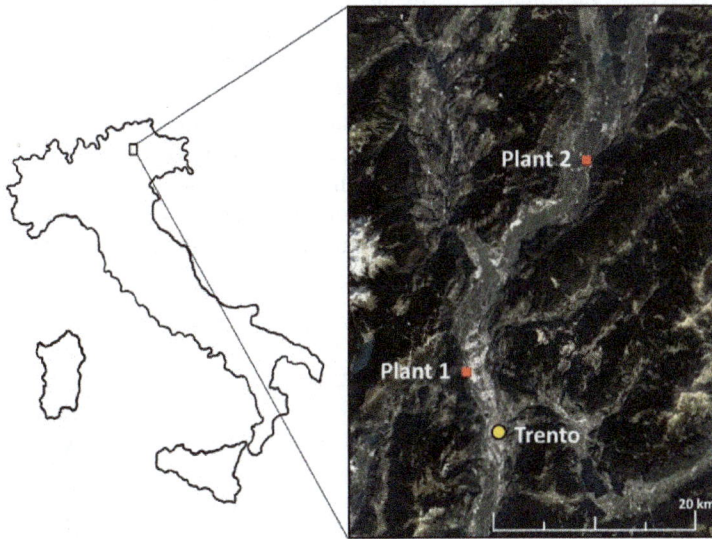

Figure 1: Locations of the Italian WtE plants selected as case studies.

The first WtE plant (Plant 1) was proposed to treat the MSW generated by the population of the province of Trento, accounting for 542,000 inhabitants [28]. The WtE plant, performing direct combustion of waste, would have treated 103,000 t/year of MSW with a nominal thermal power of 60 MW. The resulting airflow rate from the main stack, normalised to the oxygen content (11%), is 109,000 Nm^3/h. The exhaust gas would have been released from a 100 m high stack, with an outlet velocity of 20 m/s and a temperature of 140°C [29]. The stack height of 100 m was chosen based on a previous version of the project, which considered an input waste capacity of about 240,000 t/year. The same stack height was kept in the following version of the project, because lower values would have required in-depth meteorological analyses due to uncertainties in predicting the effects of a local wind (Ora del Garda), blowing from a lateral valley [30].

The second WtE plant (Plant 2) was proposed to treat 95,000 t/year of solid recovered fuel (SRF) and non-hazardous waste, with a nominal thermal power of 63 MW. The plant would have implemented the indirect combustion of waste based on waste gasification and syngas combustion. The expected airflow rate from the main stack, normalised to the oxygen content (6.4%), was 156,000 Nm^3/h. The exhaust gas would have been released by a 45-m stack, at a temperature of 130°C and with an outlet velocity of 16.5 m/s [31]. Thus, the main differences between the two plants consist in the airflow rate of the exhaust gas and in the stack height.

In the European Union (EU), the emissions from WtE plants are regulated by the Directive 2010/75/EU [32], which establishes concentration limit values for several pollutants at the emission level. The limit values concern the following air pollutants: total suspended particles (TSP), total organic carbon (TOC), hydrogen chloride (HCl), hydrogen fluoride (HF), sulphur dioxide (SO_2), nitrogen oxides (NOx), carbon monoxide (CO), cadmium (Cd) and thallium (Tl), mercury (Hg), other heavy metals, polycyclic aromatic hydrocarbons (PAHs), PCDD/Fs and dioxin-like PCBs (dl-PCBs). The other heavy metals regulated by the Directive are: antimony (Sb), arsenic (As), chromium (Cr), cobalt (Co), lead (Pb), manganese (Mn), nickel (Ni) and vanadium (V). The most critical heavy metal is Cr, which is basically composed of trivalent Cr (Cr III) and Cr VI. Differently from Cr III, Cr VI is a carcinogenic compound and its related cancer risk is dominated by the inhalation route [33]. For health risk assessment purposes, the cancer potency of chemicals is numerically quantified by the so-called slope factors (SFs), defined as the cancer risk (dimensionless) per unit of dose. A SF is expressed as the inverse of the ratio between the mass of chemical that enters the human body per body weight and unit of time, i.e. $(mg/(kg_{bw}\ d))^{-1}$. Depending on the chemical's nature, SF values can be available for inhalation (SF_{inhal}) and/or ingestion (SF_{ing}). Fig. 2 presents the SF values of the carcinogenic compounds regulated by the Directive 2010/75/EU [32].

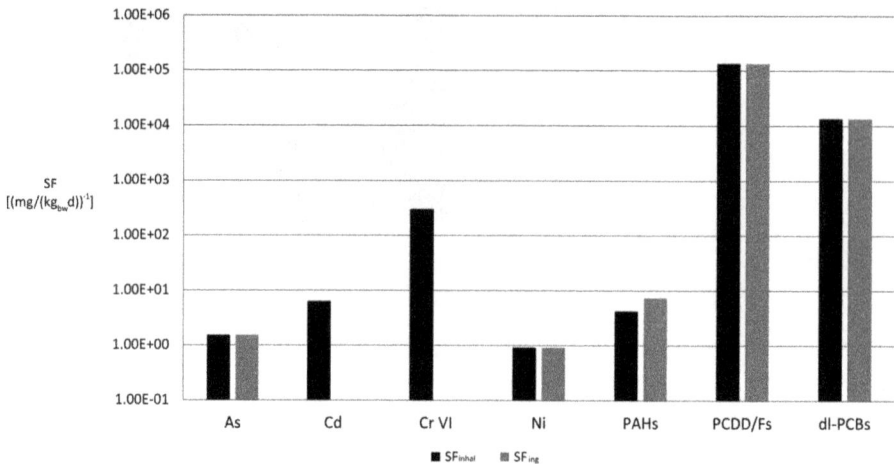

Figure 2: SF values [33] of the carcinogenic air pollutants regulated in EU.

The highest SF values are attributed to PCDD/Fs and dl-PCBs, followed by Cr VI, whose SF_{inhal} is lower than the SFs of PCDD/Fs and dl-PCBs by more than two and one orders of magnitude, respectively. As visible in Fig. 2, no SF_{ing} value has been defined for Cr VI, inhalation being its dominant exposure route. Among carcinogenic heavy metals, Cr VI is the most toxic compound, followed by Cd, As and Ni.

Table 1 presents the maximum design concentration values of the regulated air pollutants expected at the stacks of the WtE plants considered in the present paper. In the case of Cd and Tl and the large group of heavy metals regulated by the EU legislation, compound-specific values were estimated too (Table 2), following a methodology developed in a recent

Table 1: Maximum concentration values of the air pollutants regulated by the EU legislation expected at the main stacks of the two case studies [29], [31].

Air pollutant	Unit	Maximum expected stack concentration values	
		Plant 1	Plant 2
TSP	mg/Nm^3	1.5	1.5
TOC	mg/Nm^3	10	10
HCl	mg/Nm^3	2	2
HF	mg/Nm^3	0.25	0.25
SO$_2$	mg/Nm^3	10	10
NOx	mg/Nm^3	40	40
CO	mg/Nm^3	50	50
NH$_3$	mg/Nm^3	10	10
Cd+Tl	mg/Nm^3	0.025	0.025
Hg	mg/Nm^3	0.025	0.025
Other metals	mg/Nm^3	0.25	0.25
PAHs	mg/Nm^3	0.0001*	0.01
PCDD/Fs	ng$_{I-TEQ}$/Nm^3	0.03	0.025
PCBs	ng$_{I-TEQ}$/Nm^3	–	0.1

* only for benzo[a]pyrene.

Table 2: Estimation of the specific maximum stack concentrations for the heavy metals grouped according to Directive 2010/75/EU [23], [34].

Heavy metal	Estimated mass fractions in each metal group (–)	Maximum expected stack concentration (mg/Nm^3)
Cd*	0.907	0.02268
Tl	0.093	0.00233
Group total	1.00	
Sb	0.109	0.02725
As*	0.003	0.00075
Cr III	–	0.16600
Cr VI*	–	0.00500
Co	0.003	0.00075
Pb	0.075	0.01875
Mn	0.058	0.01450
Ni*	0.034	0.00850
V	0.034	0.00850
Group total	1.00	

* Carcinogenic.

paper [23]: the maximum design concentration values of the two groups (Cd+Tl and Sb, As, Cr, Co, Pb, Mn, Ni and V) was multiplied by the relative abundance of each compound in its respective group, according to the results of a metal characterisation carried out on Italian MSW incineration plants [34]. The cited study did not perform Cr speciation. However, Cr III and Cr VI concentration values were estimated considering a recent proposal for a concentration limit value for Cr VI (0.005 mg/Nm^3) [23]. The maximum concentration value expected for Cr III was calculated subtracting this value from the total Cr mass fraction in the heavy metal group (0.684) [34]. Since the maximum stack concentration values guaranteed by the two WtE plants are the same for both groups (Cd+Tl and other metals), the same metal-specific maximum concentrations are expected at both plants.

3 RESULTS AND DISCUSSION

Fig. 3 presents the approximated carcinogenic potentials of the emissions from the main stacks of Plants 1 and 2, calculated as discussed in the previous section. Concerning Plant 1, a maximum concentration value expected at the stack for dl-PCB was not provided by the proposers, thus the EU limit value was considered (0.1 ng$_{I-TEQ}$/Nm^3). Regarding PAHs, only a benzo[a]pyrene maximum design concentration value was proposed for Plant 1.

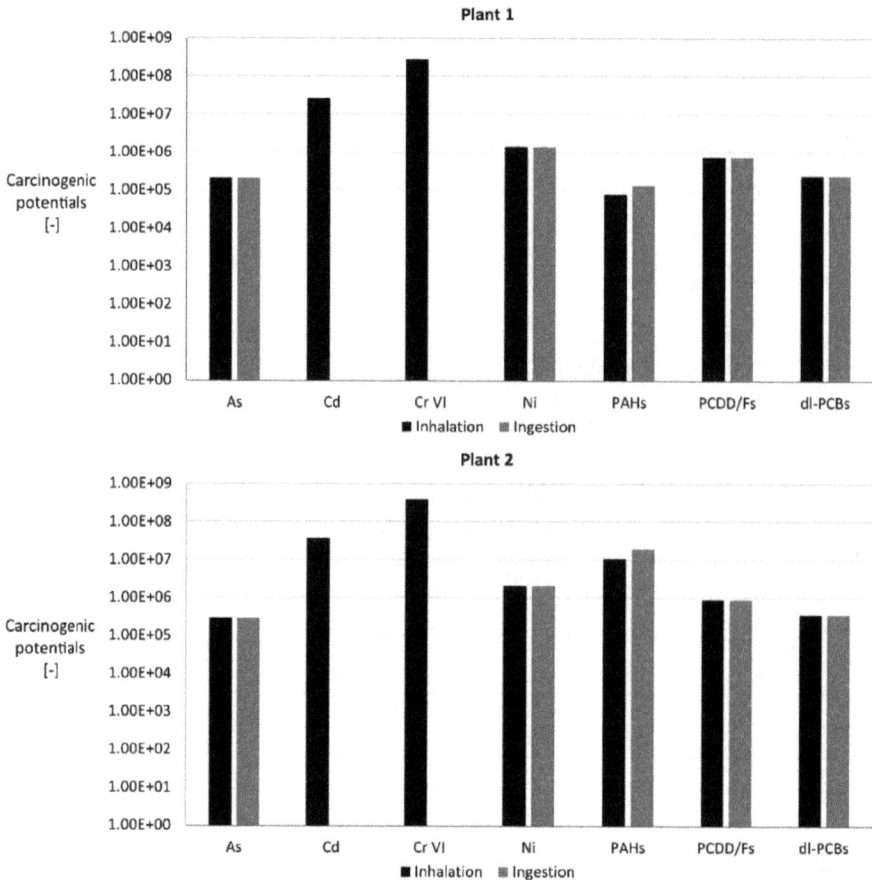

Figure 3: Carcinogenic potentials of the air pollutants released by the two WtE plants [37].

It is of course necessary to consider other key variables that may determine the fate of the air pollutants emitted (e.g., stack height, outlet velocity, exhaust gas temperature, the local meteorology, morphology, land use, the presence of population settled and its typical diet).

However, this approach can give a preliminary idea of the expected burden of carcinogenic compounds released by a plant. In addition, this approach could be adopted for other types of emission sources than WtE plants.

As shown in Fig. 3, despite the higher SFs of PCDD/Fs and dl-PCBs, the greatest contribution to the carcinogenic potentials of the mixture released at the stacks of the two WtE plants is given by two heavy metals: Cr VI and Cd, the first giving a 10-times higher contribution than the second. This is the result of the stack concentration limit values set by the legislation, which defines single total values for Cd+Tl and the sum of the other heavy metals including Cr VI.

In fact, each group is dominated by a single compound: Cd, in the Cd+Tl group, and Cr VI in the other group. The presence of single concentration limit values for groups of heavy metals led the proposers of the plant to define single maximum concentration values guaranteed by the plants for the respective groups. It is worth noting that, according to this legislative scheme, the limit value set for the largest group of heavy metals is 0.5 mg/Nm^3. Here, a 100-times lower concentration value was adopted for Cr VI, because of its carcinogenicity. From Table 2, it is possible to calculate the percentage of Cr VI in total Cr (Cr III + Cr VI), which is 2.9%. This percentage is in the range of values available in the few literature studies on Cr speciation in the exhaust gas of WtE plants, which has been found to be 1.15%–6.5% [35], [36]. Thus, Cr VI is likely to be the dominant pollutant for the inhalation exposure route, followed by Cd. On the other hand, Ni and persistent organic pollutants (PCDD/Fs, dl-PCBs and PAHs) are likely to be the dominant pollutants for the exposure via diet. Therefore, in contexts with limited agriculture and/or livestock activities, i.e. where the consumption of local food products is limited, health risk assessment procedures should pay specific attention to Cr VI and Cd. Reducing the release of heavy metals thus becomes a priority.

Besides the higher carcinogenic potentials of Cr VI compared to other pollutants, this compound has proved to be difficult to measure in the environment. According to the feasibility study of Plant 1 [29], the Cr VI detection limit in total suspended particle samplings in ambient air is 2 ng/m^3. The continuative inhalation exposure to such Cr VI concentration would induce an excess cancer risk of $2.4 \cdot 10^{-5}$ [37], i.e., 24-times higher than the acceptable value for the exposure to single contaminants defined by the Italian legislation [38]. Thus, the compliance with this limit value could not be verified by field monitoring. Dispersion modelling would be the only applicable methodology. However, an EIA should produce/select local meteorological data with particular care to characterise the impact area accurately. As a matter of fact, the dispersion of air pollutants from a source located in a mountainous region is generally weaker compared to coastal areas, uplands or flatlands. This is visible by a simple comparison between the so-called dilution factors (DFs) of an emission source located in a mountainous region and a similar source located in other geographical contexts. DFs are defined as the ratio between the highest value of the average ambient air concentration (or deposition to soil) of a pollutant on the computational domain chosen for the dispersion simulations required in the EIA process and its respective mass flowrate at the emission level. DFs provide an estimate of the maximum impacts of the different pollutants expected at ground level. The higher the DF, the higher is the impact estimated at ground level. To estimate the impacts of the single heavy metals belonging to a group of metals according to the legislation (e.g., Cr VI in the heavy metal group considered by the EU legislation), it is reasonable to consider the DF of TSP, since metals are mainly released in particle phase [23]. As an example, based on the results of the dispersion simulations carried out during the EIA process, the DF of TSP ambient air concentration referred to Plant 1 is $8.2 \cdot 10^{-6}$ s/m^3 [29]. According to a survey on MSW incinerators in Italy [39], the DF of a

WtE plant located near the coast of the Adriatic Sea is $2.7 \cdot 10^{-7}$ s/m^3, thus 30 times lower than Plant 1.

On the side of emission prevention, the release of Cr VI can be reduced by adopting improved air pollution control technologies, based on wet or dry processes. Cr VI shows a high solubility in water, thus adopting wet processes (with further wastewater treatment) could be a viable option. However, the tendency of the WtE sector is to prefer dry technologies for ease of management. Hence the importance of proposing a specific limit value for Cr VI and a routine monitoring of its stack concentration. Once the role of Cr VI will have been verified (e.g., via the carcinogenic potential methodology or, more thoroughly, via a complete health risk assessment), it is appropriate to consider setting a dedicated concentration limit value at the stack. Once this in-depth study on Cr VI is performed, it is desirable to implement a similar approach for Cd.

In case the inhalation route of exposure proved to induce an acceptable excess cancer risk, the health risk assessment should focus on the potential dietary intake of pollutants by the local population. It is crucial to both appropriately select the food crops and livestock activities present in the area and retrieve detailed statistics on the local food consumption and dietary habits. The traditional health risk assessment methodology for the ingestion of contaminated food [40] should then be applied, also considering food-chain models available in the literature in case some food products lack specific formulations.

In unfavourable situations like mountainous regions, it is worth considering alternative WtE approaches. One option is to subdivide a large combustion plant into more smaller plants placed in different locations, in order to lower the local impacts. However, this option entails higher investment costs that might not be sustainable for the local government and community. A more economically and environmentally sustainable opportunity is given by waste gasification: instead of burning the syngas generated in the gasification stage, it is possible to convert it into chemicals (e.g., methanol, ethanol, hydrogen, dimethyl ether) that may also partly replace fossil fuels, also preventing odour problems. This way, both local and global environmental impacts would be reduced, as indicated also in recent literature [41], [42].

4 CONCLUSIONS

The present paper is a contribution to solutions to the environmental problems induced by large combustion plants in mountainous regions, like the Alps. To better visualise the potential implications involved, two case studied regarding two waste combustion plants were discussed. A quick method to estimate the carcinogenic potentials of the pollutants released from a source was provided. This method helps define which pollutants should receive priority in a health risk assessment, thus reducing the risk of neglecting key pollutants. The application of a simplified methodology based on the DFs of TSP, which must be calculated from the results of dispersion simulations, may provide indications on the impact expected from single contaminants at ground level and on the benefits achievable when setting specific emission limit values. The present paper also provided insights into a proper application of health risk assessment procedures according to the results of the screening based on the carcinogenic potentials of the pollutants emitted from a plant, on the land use and local diet of the exposed population. Finally, an unconventional WtE scheme was proposed to reduce the local impacts of WtE plants based on combustion. Such scheme, based on waste gasification and ex-situ valorisation of syngas, is preferable in situations with unfavourable morphology, relatively high amount of input waste and where the construction of multiple small-size waste combustion plants is not economically feasible.

REFERENCES

[1] Anquetin, S., Guilbauld, C. & Chollet, J.-P., Thermal valley inversion impact on the dispersion of a passive pollutant in a complex mountainous area. *Atmospheric Environment*, **33**, pp. 3953–3959, 1999.

[2] Triantafyllou, A.G. & Kassomenos, P.A., Aspects of atmospheric flow and dispersion of air pollutants in a mountainous basin. *Science of the Total Environment*, **297**, pp. 85–103, 2002.

[3] Boanini, C., Mecca, D., Pognant, F., Bo, M. & Clerico, M., Integrated mobile laboratory for air pollution assessment: Literature review and cc-trairer design. *Atmosphere,* **12**(8), p. 1004, 2021.

[4] Falocchi, M., Zardi, D. & Giovannini, L., Meteorological normalization of NO_2 concentrations in the Province of Bolzano (Italian Alps). *Atmospheric Environment*, **246**, 118048, 2021.

[5] Quimbayo-Duarte, J., Chemel, C., Staquet, C., Troude, F. & Arduini, G., Drivers of severe air pollution events in a deep valley during wintertime: A case study from the Arve river valley, France. *Atmospheric Environment*, **247**, 118030, 2021.

[6] Rada, E.C., Ragazzi, M. & Malloci, E., Role of levoglucosan as a tracer of wood combustion in an alpine region. *Environmental Technology*, **33**(9), pp. 989–994, 2012.

[7] Schiavon, M., Antonacci, G., Rada, E.C., Ragazzi, M. & Zardi, D., Modelling human exposure to air pollutants in an urban area. *Revista de Chimie,* **65**(1), pp. 61–64, 2014.

[8] Torretta, V., Rada, E.C., Panaitescu, V. & Apostol, T., Some considerations on particulate generated by traffic. *UPB Scientific Bulletin, Series D: Mechanical Engineering*, **74**(4), pp. 241–248, 2021.

[9] Pietrogrande, M.C., Bertoli, I., Clauser, G., Dalpiaz, C., Dell'Anna, R., Lazzeri, P., Lenzi, W. & Russo, M., Chemical composition and oxidative potential of atmospheric particles heavily impacted by residential wood burning in the alpine region of northern Italy. *Atmospheric Environment*, **253**, 118360, 2021.

[10] Schmidl, C., Marr, I.L., Caseiro, A., Kotianova, P., Berner, A., Bauer, H., Kasper-Giebl, A. & Puxbaum, H., Chemical characterisation of fine particle emissions from wood stove combustion of common woods growing in mid-European Alpine regions. *Atmospheric Environment,* **42**(1), pp. 126–141, 2008.

[11] Nanni, A., Brusasca, G., Calori, G., Finardi, S., Tinarelli, G., Zublena, M., Agnesod, G. & Pession, G., Integrated assessment of traffic impact in an Alpine region. *Science of the Total Environment*, **334–335**, pp. 465–471, 2004.

[12] Adami, L., Schiavon, M. & Rada, E.C., Potential environmental benefits of direct electric heating powered by waste-to-energy processes as a replacement of solid-fuel combustion in semi-rural and remote areas. *Science of the Total Environment*, **740**, 140078, 2020.

[13] Ozgen, S., Ozgen, S., Caserini, S., Galante, S., Giugliano, M., Angelino, E., Marongiu, A., Hugony, F., Migliavacca, G. & Morreale, C., Emission factors from small scale appliances burning wood and pellets. *Atmospheric Environment*, **9**, pp. 144–153, 2014.

[14] Fadel, M., Ledoux, F., Seigneur, M., Oikonomou, K., Sciare, J., Courcot, D. & Afif, C., Chemical profiles of $PM_{2.5}$ emitted from various anthropogenic sources of the Eastern Mediterranean: Cooking, wood burning, and diesel generators. *Environmental Research*, **211**, 113032, 2022.

[15] Directive 2008/98/EC of the European Parliament and of the Council of 19 November 2008 on waste and repealing certain Directives; EUR-Lex. https://eur-lex.europa.eu/legal-content/EN/TXT/HTML/?uri=CELEX:02008L0098-20180705. Accessed on: 29 Mar. 2022.

[16] Hou, C. & Sarigöllü, E., Waste prevention by consumers' product redistribution: Perceived value, waste minimization attitude and redistribution behavior, *Waste Management,* **132,** pp. 12–22, 2021.

[17] Wilts, H., Fecke, M. & Zeher, C., Economics of waste prevention: Second-hand products in Germany. *Economies, 9*(2), p. 74, 2021.

[18] Cappellaro, F., Fantin, V., Barberio, G. & Cutaia, L., Circular economy good practices supporting waste prevention: The case of Emilia-Romagna Region. *Environmental Engineering and Management Journal,* **19**(10), pp. 1701–1710, 2020.

[19] Kurniawan, T.A., Lo, W., Singh, D., Othman, M.H.D., Avtar, R., Hwang, G.H., Albadarin, A.B., Kern, A.O. & Shirazian, S., A societal transition of MSW management in Xiamen (China) toward a circular economy through integrated waste recycling and technological digitization. *Environmental Pollution,* **277,** 116741, 2021.

[20] Klinghoffer, N.B. & Castaldi, M.J., *Waste to Energy Conversion Technology,* Woodhead Publishing, pp. 15–28, 2013.

[21] Zubero, M.B., Aurrekoetxe, J.J., Ibarluzea, J.M., Rivera, J., Parera, J., Abad, E., Rodríguez, C. & Sáenz, J.R., Evolution of PCDD/Fs and dioxin-like PCBs in the general adult population living close to a MSW incinerator. *Science of The Total Environment,* **410–411,** pp. 241–247, 2011.

[22] García, F., Barbería, E., Torralba, P., Landin, I., Laguna, C., Marquès, M., Nadal, M. & Domingo, J.L., Decreasing temporal trends of polychlorinated dibenzo-p-dioxins and dibenzofurans in adipose tissue from residents near a hazardous waste incinerator. *Science of The Total Environment,* **751,** 141844, 2021.

[23] Rada, E.C., Schiavon, M. & Torretta, V., A regulatory strategy for the emission control of hexavalent chromium from waste-to-energy plants. *Journal of Cleaner Production,* **278,** 123415, 2021.

[24] Zheng, J., Yu, L., Ma, G., Mi, H. & Jiao, Y., Residents' acceptance towards waste-to-energy facilities: Formation, diffusion and policy implications. *Journal of Cleaner Production,* **287,** 125560, 2021.

[25] National Environmental Policy Act (NEPA), https://ceq.doe.gov/index.html. Accessed on: 30 Mar. 2022.

[26] Schiavon, M., Rada, E.C., Adami, L., Fox, F. & Ragazzi, M., Integrated methodology for the management of human exposure to air pollutants. *WIT Transactions on Ecology and the Environment,* vol. 236, WIT Press: Southampton and Boston, pp. 287–296, 2019.

[27] Liu, Y., Xu, M., Ge, Y., Cui, C., Xi, B. & Skitmore, M., Influences of environmental impact assessment on public acceptance of waste-to-energy incineration projects. *Journal of Cleaner Production,* **304,** 127062, 2021.

[28] Statistiche ISTAT, Istituto Italiano di Statistica. http://dati.istat.it/. Accessed on: 31 Mar. 2022.

[29] Barbone, F., Brevi, F., Ghezzi, U., Ragazzi, M. & Ventura, A., Concessione di lavori per la progettazione, realizzazione e gestione dell'impianto di combustione o altro trattamento termico con recupero energetico per rifiuti urbani e speciali assimilabili in località Ischia Podetti, nel Comune di Trento – Studio di fattibilità. Provincia Autonoma di Trento, 2009.

[30] Laiti, L., Zardi, D., de Franceschi, M. & Rampanelli, G., Atmospheric boundary layer structures associated with the Ora del Garda wind in the Alps as revealed from airborne and surface measurements. *Atmospheric Research,* **132–133,** pp. 473–489, 2013.

[31]   Archive of Environmental Assessment Procedures, Environmental Protection Agency of the Province of Bolzano. https://ambiente.provincia.bz.it/valutazioni-ambientali/archivio-procedure-via-vas-screening-aia.asp. Accessed on: 31 Mar. 2022.

[32]   Directive 2010/75/EU of the European Parliament and of the Council of 24 November 2010 on industrial emissions (integrated pollution prevention and control). https://eur-lex.europa.eu/legal-content/EN/TXT/?uri=celex%3A32010L0075. Accessed on: 31 Mar. 2022

[33]   United States Environmental Protection Agency, Chemical specific parameters November 2021. https://semspub.epa.gov/work/HQ/401667.pdf. Accessed on: 4 Apr. 2022.

[34]   ARPA Emilia-Romagna, Le emissioni degli inceneritori di ultima generazione. https://www.arpae.it/it/documenti/pubblicazioni/le-emissioni-degli-inceneritori-di-ultima-generazione. Accessed on: 4 Apr. 2022.

[35]   UK Environment Agency, Releases from municipal waste incinerators: Guidance to applicants on impact assessment for group 3 metals stack. https://www.gov.uk/government/uploads/system/uploads/attachment_data/file/532474/LIT_7349.pdf. Accessed on: 5 Apr. 2022.

[36]   Świetlik, R., Trojanowska, M., Łożyńska, M. & Molik, A., Impact of solid fuel combustion technology on valence speciation of chromium in fly ash. *Fuel*, **137**, pp. 306–312, 2014.

[37]   United States Environmental Protection Agency, Chromium(VI) CASRN 18540-29-9 DTXSID7023982. https://iris.epa.gov/ChemicalLanding/&substance_nmbr=144. Accessed on: 5 Apr. 2022.

[38]   Repubblica Italiana, Ulteriori disposizioni correttive ed integrative del decreto legislativo 3 aprile 2006, n. 152, recante norme in materia ambientale. http://www.camera.it/parlam/leggi/deleghe/08004dl.htm. Accessed on: 5 Apr. 2022.

[39]   Regione Emilia-Romagna, Emissioni degli inceneritori e modelli di ricaduta. https://www.arpae.it/it/documenti/pubblicazioni/i-quaderni-di-moniter. Accessed on: 6 Apr. 2022.

[40]   United States Environmental Protection Agency, Human health risk assessment protocol for hazardous waste combustion facilities. https://nepis.epa.gov/Exe/ZyPDF.cgi/P10067PR.PDF?Dockey=P10067PR.PDF. Accessed on: 5 Apr. 2022.

[41]   Tubino, M., Adami, L. & Schiavon, M., Environmental impact of waste-to-energy processes in mountainous areas: the case of an Alpine region. *11th International Conference on Waste Management and Environmental and Economic Impact on Sustainable Development*, 2022.

[42]   Adami, L., Schiavon, M. & Tubino, M., An in-depth analysis on odour dispersion modelling and its application to waste management operations. *11th International Conference on Waste Management and Environmental and Economic Impact on Sustainable Development*, 2022.

ENVIRONMENTAL IMPACT EVALUATION OF ODOR DISPERSION EMITTED FROM PIG FARMS: A CASE STUDY FROM ALBAN, CUNDINAMARCA, COLOMBIA

GERARDO ROMERO-TOVAR[1], HERNAN D. GRANDA-RODRIGUEZ[2,3]
& MIGUEL A. DE LUQUE-VILLA[2,3]
[1]Corporación Autónoma Regional de Cundinamarca CAR, Colombia
[2]Grupo de Investigación Cundinamarca Agroambiental, Facultad de Ciencias agropecuarias,
Universidad de Cundinamarca, Colombia
[3]Departamento de Ecología y Territorio, Facultad de Estudios Ambientales y Rurales,
Pontificia Universidad Javeriana, Colombia

ABSTRACT

The purpose of this research was to determinate the environmental impacts generated by the offensive odors coming from pig farming in the Alban municipality, located in Cundinamarca, Colombia. The main objective was to propose guidelines for the elaboration of a Plan for Offensive Odor Impact Reduction for the pig farms of the municipality. Initially an environmental impact evaluation was carried out. Then the application of the protocols established in Colombian regulations: NTC 6012-1 and Resolution 1543 of 2013 of the Ministry of Environment and Sustainable Development (MADS) resulted in a relevant impact on the air quality. This was ratified in the developed psychometric analysis that allowed us to identify a particular nuisance of offensive odors in the village of Pantanillo within the sector located on the right side of the road that communicates Albán and Guayabal de Siquima, taking into account that among the results obtained within the psychometric analysis the perception by odor level showed values between "Strong" (35.0%) and "Very strong", (20.0%) for the possible affected zone, while in the control area C, 72% considered it "very faint" and "no odor", finally proposing guidelines for the development of a program of reduction of offensive odors that fits the type of population existing in the study area and allow a reduction in the discomfort by offensive odors.
Keywords: environmental impacts, offensive odor, pig farming, reduction of offensive odors.

1 INTRODUCTION

Environmental odors represent a special air pollution problem that sometimes overrides the general air quality [1]. Odor-producing compounds like ammonia, aliphatic amines, hydrogen sulfide, carbon disulfide, mercaptans, BTEX, chlorobenzenes and chloroform originate from various activities like waste management plants, industrial production, and animal production facilities [2]. Odor from pig styes or pig production is a severe problem that causes substantial nuisance to neighbors and prevents the farmers from developing production [3], [4]. Ammonia emissions from intensive pig production are a major public concern due to their potential effects as a public nuisance [5].

In Colombia, pig farming is one of the economic sectors with the greatest growth in recent decades. Despite the world economic situation having harmful effects on the national economy, national pork production reached 46,888,080 tons [6], However, its production generates environmental impacts on water, soil and air. One of the main impacts of the sector is the generation of odors. Although such odors are not always harmful to people's health, they do generate nuisances for the communities surrounding pig husbandry. This type of nuisance causes displeasure and discomfort for the surrounding community and generates environmental impacts on the air quality. This affects the quality of life because it makes it uncomfortable to eat and sleep, and interrupts the daily life of the people who generally live

WIT Transactions on Ecology and the Environment, Vol 259, © 2022 WIT Press
www.witpress.com, ISSN 1743-3541 (on-line)
doi:10.2495/AWP220061

in the immediate vicinity of the pig farms. Because of the above, it is necessary to look at this line of the economy that has great potential for improvement and development, hence the importance of proposing good practices for the management of offensive odors in this sector.

Quantitative and qualitative characterization of odors can be carried out by direct or indirect methods of odor control, and the following approaches are used: dynamic olfactometry, dispersion models, field inspection, electronic noses, and odor surveys [7]. The Resolution 1541 of 2013 [8] in Colombia, establishes ambient air quality standards and source emission assessment of offensive odors. The application of this resolution establishes the evaluation of a complaint using standardized surveys.

Colombian regulations establish permissible levels of air quality or emissions, for the case of offensive odors, and the standard peaks of reference conditions at 25°C and 760 mm Hg. Additionally, it provides procedures for the evaluation of activities that generate offensive odors, and it establishes measurement methods by analytical techniques and the prevention of odor generation through the odors impact reduction plans (OIRP). This paper evaluated a psychometric analysis based on the NTC 6012, for knowing the magnitude of the nuisance associated with offensive odors from this activity and generated guidelines for odors impacting reduction plans (OIRP).

2 ODOR'S ENVIRONMENTAL IMPACT

Air pollution from odor compounds is a significant problem for cities nowadays Odor emissions are a common source of complaints, affecting the quality of life for people. Odor is a property of a mixture of different volatile chemical species (sulfur, nitrogen, and volatile organic compounds) capable of stimulating olfaction sufficiently to trigger the sensation of odor [7]. Exposure to environmental odors is one of the major causes of complaints made by residents living near different kinds of industrial and agricultural settlements [12]. The harmful effects of odors are not related to their toxic effects on the body but result from people´s subjective reception and evaluation that has an adverse effect on the human psyche in the long run [13].

The most important pollutants that impact the environment that are emitted from livestock buildings used for the production of monogastric animals are odors and ammonia (NH_3). Odorous substances are more relevant on a local scale, causing annoyance to nearby residents [14]. The highest rate of odor emissions came mostly from pig farms [15] and represent one the most current topics in terms of industrial pig farming pollution effects, especially because of nearby settlements [16].

One of the first research projects found is the one carried out in the United States by Douglas Kreis [17] in which the author indicated that the main complaints of animal production industries in the United States were caused by odors emanating from them. He mentions in his research that controls to reduce the impact of offensive odors were costly and limited to the generation or quality of animal production, and he proposed land use planning and zoning for agricultural/animal feeding purposes as a tool to reduce offensive odors.

Nicell [18] carried out an evaluation of the environmental impact of odors and their regulation, measuring the impact of offensive odors according to measurable and objective criteria. In the existing regulation of odors, they took into account the annoyances generated by people according to their perception. The author indicated that the measurement of odors comes from a series of variables known as frequency, intensity, duration, offensive, and location, proposing an approach based on these variables.

At the Latin American level, important investigations were found, such as that of Murguía [19], where the economic repercussions of production and the nature of nuisances due to offensive odors were discussed, as well as an analysis of the current state of Mexican legislation for the control and regulation of these odors. He initially indicated how the perception of odors can affect human senses and how they can compromise people's quality of life. He also proposed to legislate the impact of odors considering several variables, among them, the distance from the company that emits the odor, size, type of company, and the manufacturing practices of each one. He also mentioned the need to take into account two types of measurement, in any in-situ odor study using an atmospheric dispersion model, and a complementary one, through legislation based on complaints.

In Ecuador [20] an analysis of offensive odors was carried out and mitigation proposals were proposed for an area in Guayaquil. The methodology used for the analysis is the passive measurement of H_2S, sectoring the place with a measurement range of 0.2 to 200 ppm within 48 hours. It was concluded that the measurements exceed the threshold established at the international level, so it was proposed to specify the thresholds for the emission of offensive odors in the city, since they were not defined.

In most countries, environmental legislation covers most types of common air pollutants; and there is little variation between jurisdictions with such legislation. However, odor legislation tends to be much more varied and varies across a wide spectrum: from having little to no specific mentioning in environmental legislation to extensive and rigid detailing in odor source testing, odor dispersion modelling, ambient odor monitoring, setback distances, process operations, and odor control procedures. Odor legislation can be highly variable from one jurisdiction to the next [21].

3 MATERIALS AND METHODS

3.1 Study area

The possible impact area of this study is located in the Chimbe and Pantanillo villages of the Albán municipality in Cundinamarca province. The place where the activity took place was a rural area near the town, at the coordinates 4°54'49.78"N, 74°28'00.83"W (Fig. 1).

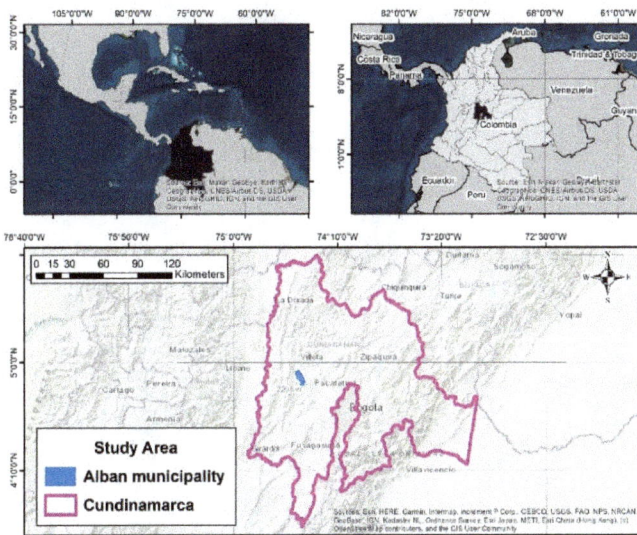

Figure 1: Study area – Alban, Cundinamarca.

3.2 Nuisance odor evaluation

This evaluation was carried out according to the protocols established in NTC 6012-1 and the guidelines required by resolution 1541 of 2013 [8] and the Protocol adopted by resolution 2087 of 2014 [22] as shown in Fig. 2.

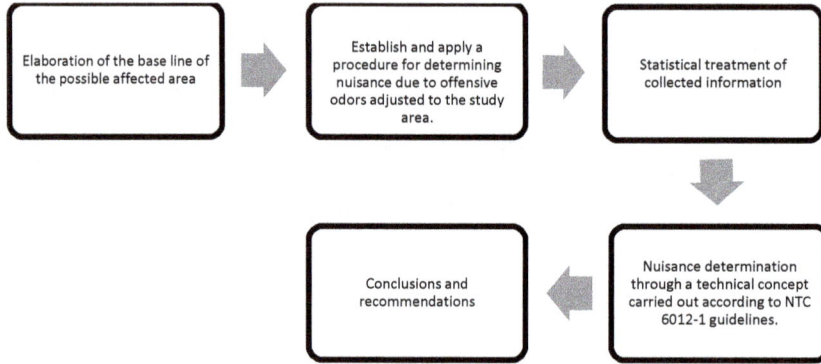

Figure 2: NTC 6012-1 Nuisance odor evaluation.

NTC 6012-1 standards were followed, selecting a possible effective zone (A) and a control zone (C) that must share similar characteristics of environmental, geographic, socioeconomic, housing infrastructure, and vehicular flow. Also, wind direction was taken into account. The data was retrieved from the Colombian "Instituto de Hidrología, Meteorología y Estudios Ambientales – IDEAM"; the nearest weather station was "Tibaitata [21205420]"; the data analysis period was 2017–2021 (Fig. 3). Therefore, in compliance with the technical standards, the possible impact area (A) was defined in the Chimbe village and the control area (C) in Pantanillo village in Albán (Fig. 4).

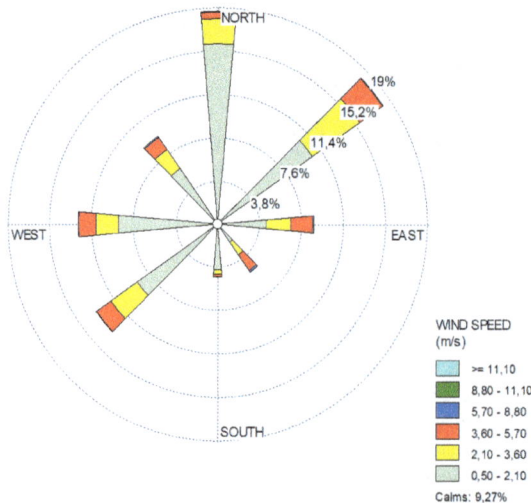

Figure 3: Wind rose study area.

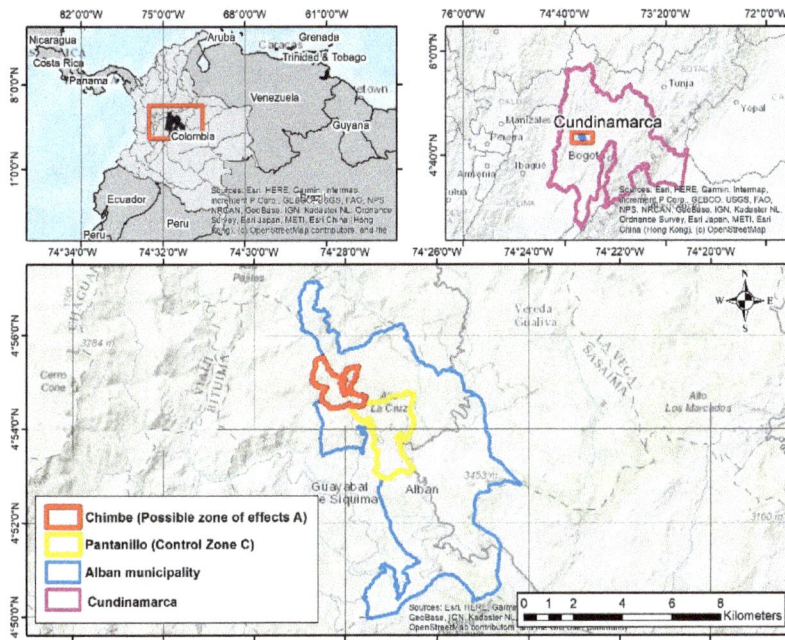

Figure 4: Possible impact area and control zone.

A psychometric analysis through the application of the surveys were applied to people older than 18 years in the possible zone of effects (A) and a control zone (C). The sample size was calculated by the SurveyMonkey sample size calculator [23] with a confidence level of 95% and a margin of error at 5%. The calculated sample size was 11 for Pantanillo and 20 for Chimbe.

The first section of the survey made it possible to characterize the people surveyed based on their spatial location in the area and sociodemographic conditions such as age, economic activity of the property, and how long they have lived in the area. The second part of the instrument presented the following questions: (1) How strong do you perceive odors in the area? (2) How often do you perceive odors in the area? (3) How would you rate the annoyance due to odor in the area where your property is located? (4) Do you consider that discomfort in this area is tolerable or intolerable?

To assess whether or not there were differences in the perception of odors between the control area and the possible affected area, a chi-square test of homogeneity ($\chi 2$) was performed. The results were obtained with the chisq.test package for the R platform [24].

4 RESULTS

4.1 Odor perception

According to those interviewed in the possible zone of effect (A), we found that 35.0% of the population considered that the level of odor was "Strong", 25.0% "Very faint odor", 20% as "Very strong", 10% "Faint odor", and finally 5% as "Distinct odor" and "No odor". While for the control zone (C) the results were that the perception with the highest number of responses was "no odor" 36% and "very faint odor" 36%, followed by 27%, as "faint odor", as shown in Fig. 5. The results show that there were statistical differences in the perceptions of people between zone A and the control zone C ($\chi2$=12.557, df = 5, p-value = 0.02791) where these odors were perceived much more in the affected zone.

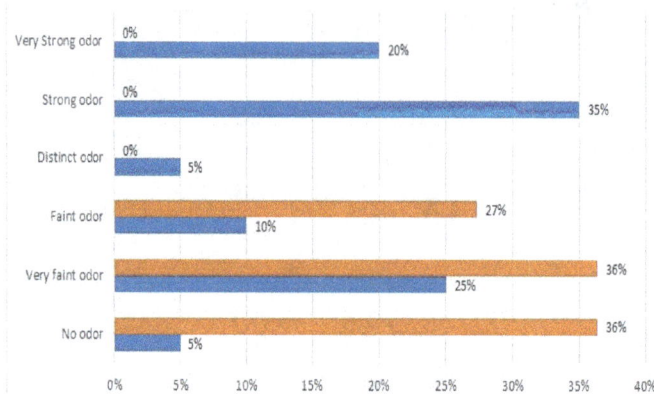

Figure 5: Odor perception.

4.2 Odor perception frequency

The relative frequencies of the data were calculated for the possible effect zone (A), we found that 40% have perceived the odors "Everyday", 25% "2 or 3 times a week", 15.0% "2 or 3 times a month", 10% "1 day a week" and 5% "1 day a month". The control zone (C) showed the odor frequencies that predominate are "Never" and "1 day a month" with 36% and 27% respectively, followed in equal proportion with 9%, the categories "2 or 3 times a month", "1 day a month", "2 or 3 times a week" and "Everyday" (Fig. 6).

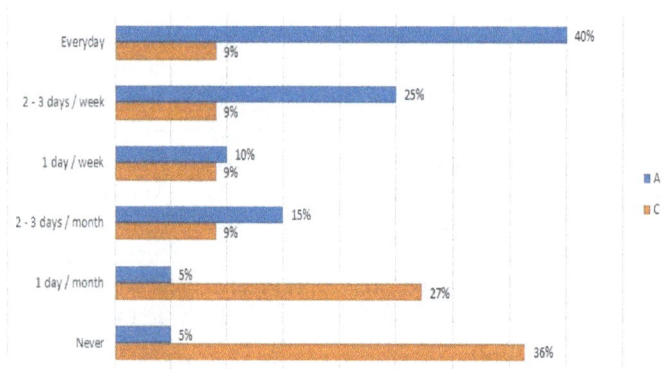

Figure 6: Odor perception frequency.

4.3 Odor nuisance levels

The responses obtained in a possible effect zone (A) for nuisance levels due to odors was level 10 with 35%, then level 9 with 15%, the same as level 5, followed by level 8, 7, and 0 with 10% and finally level 4 with 5%. For the control zone (C) nuisance levels due to odors were level 1 with 45%, followed by level 3 with 18%, and levels 0, 1, 4 and 8 with 9%, as shown in Fig. 7.

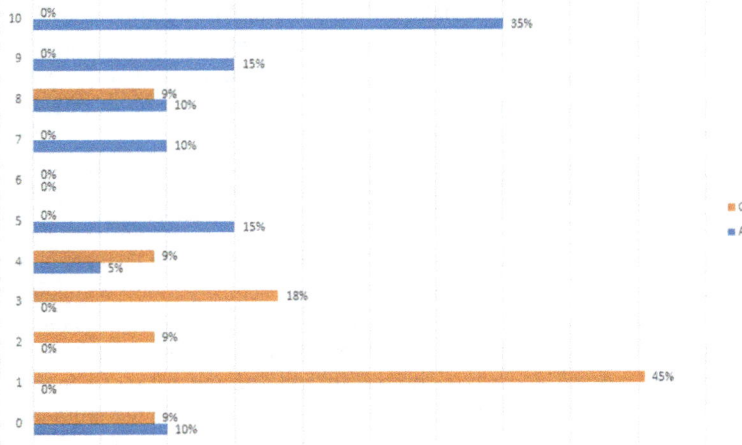

Figure 7: Odor nuisance levels.

4.4 Nuisance tolerance

For the possible zone of effect (A), 60% of the population considered the situation intolerable, and the remaining 40% say it was tolerable, while for control zone C, 100% of the population considered the odor nuisance tolerable.

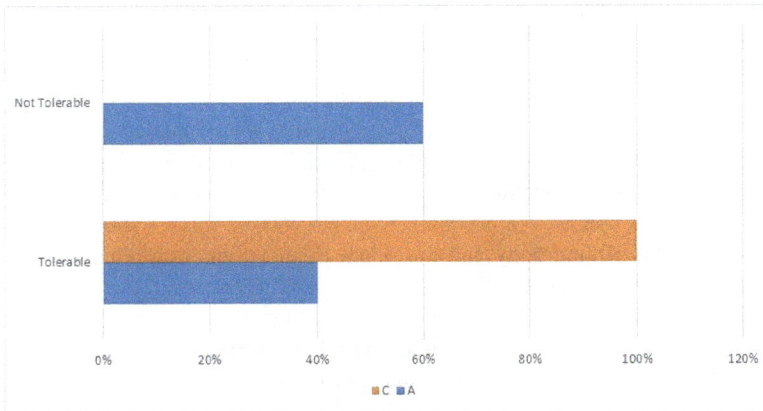

Figure 8: Nuisance tolerance.

4.5 Odors impact reduction plans (OIRP)

In the villages of Pantanillo and Chimbe selected as the study area the productive units that carry out the pig farm activity house a range of 1 to 400 pigs. According to the analysis, this is an area of low economic income, so they will work with good practices as the main method of reduction, identifying that the main problem is odors [6]. For this, offensive odor reduction programs relating to the livestock sector were reviewed. We identified strong deficiencies in the social area that generated various effects divided between the effect of technical activities, whether good practices and/or best available techniques, and the perception of those possibly affected by offensive odors. Individual perception in many cases eliminates any positive effect on the community of the actions implemented by the possible odor generator [25]. Proposed techniques aim to prevent and reduce emissions of gases by intervening in the processes responsible for its formation and volatilization.

Initially, the installation of natural barriers was proposed that contributed to mitigate impacts and reduce the dispersal of odors that are inherent in production that can affect the communities near the pig farms [26]. Additionally, these barriers improved appreciation of the surrounding populations and the influence of subjective aspects in the perception of odors. Taking into account that according to where the wind arose, the general wind direction in the Albán municipality is north and northeast, this is the direction in which the planting of species should be prioritized that will be arranged approximately 20 m from the boundary of each unit. Planting should not be done at a shorter distance because it can generate an effect contrary to the expected effect, limiting the circulation of air inside the unit and giving rise to different odors. In addition to this condition, it should also be taken into account that these barriers will be formed mainly by three rows of trees. These trees should be scaled in height so that the first row is formed by shrubs, the second by species of medium height, and the third by deciduous trees, orienting the row of shrubs towards the odor-producing units and the taller trees towards the neighboring boundaries.

Reducing the temperature inside the housing and the airflow over the surface of the pig farm can reduce ammonia emissions. But it must be taken into account that guaranteeing an adequate temperature and air renewal are two essential premises both for the welfare of the animals and for the maintenance of their productive yields. Therefore, ventilation and air conditioning systems must always be adjusted according to the needs and comfort of the animals. The following management indicators are presented for the internal control of the air and the conditioning of the housing.

5 CONCLUSIONS

It was possible to identify that conflicts are generated by offensive odors due to poor operating practices. After reviewing existing regulations in Colombia, we determined that the techniques to be developed for the assessment of offensive odors could be validated through a survey in the sector.

Once reviewed the bibliography and the odors impact reduction plans (OIRP). So, we determined that in medium and small producers, typical producers of the study area did not have the technical expertise nor the financial capacity to carry out a OIRP.

It is necessary to use specific guidelines based on good technical practices that are adjusted to the budgets of medium and small producers, as well as an awareness of the activities and investments aimed at odor reduction known by the affected population and so allow changing the perceptions of offensive odors in the study area.

The results of these studies are important for the scientific community because there are very few studies in Latin America and Colombia on the perception of odors in communities near pig farms. This type of study allows assessment whether a community is being affected by offensive odors without the need for monitoring.

ACKNOWLEDGEMENT
The authors thank the University of Cundinamarca for all of their support and information they provided for the project.

REFERENCES

[1] Chunrong, J. et al., Identification of origins and influencing factors of environmental odor episodes using trajectory and proximity analyses. *J. Environ. Manage.*, **295**, 2021. DOI: 10.1016/j.jenvman.2021.113084.

[2] M.Piccardo, M., Geretto, M., Pulliero, A. & Izzotti, A., Odor emissions: A public health concern for health risk perception. *Environ. Res.*, 135907, 2021. DOI: 10.1016/j.envres.2021.112121.

[3] Hansen, M.J., Jonassen, K.E.N., Løkke, M.M., Adamsen, A.P.S. & Feilberg, A., Multivariate prediction of odor from pig production based on in-situ measurement of odorants. *Atmos. Environ.*, **135**, pp. 50–58, 2016. DOI: 10.1016/j.atmosenv.2016.03.060.

[4] Hove, N.C.Y., Van Langenhove, H., Van Weyenberg, S. & Demeyer, P., Comparative odour measurements according to EN 13725 using pig house odour and n-butanol reference gas. *Biosyst. Eng.*, **143**, pp. 119–127, 2016. DOI: 10.1016/j.biosystemseng.2016.01.002.

[5] Tabase, R.K. et al., Effect of ventilation control settings on ammonia and odour emissions from a pig rearing building. *Biosyst. Eng.*, **192**, pp. 215–231, 2020. DOI: 10.1016/j.biosystemseng.2020.01.022.

[6] PorkColombia, Economía porcícola 2020. *Rev. PorkColombia*, **257**, p. 60, 2021.

[7] Conti, C., Guarino, M. & Bacenetti, J., Measurements techniques and models to assess odor annoyance: A review. *Environ. Int.*, **134**, 105261, 2020. DOI: 10.1016/j.envint.2019.105261.

[8] Ministerio de Ambiente & Desarrollo Sostenible, Resolución 1541 del 12 de noviembre de 2013. Bogotá, DC, Colombia, p. 13, 2013.

[9] Brancher, M., Griffiths, K.D., Franco, D. & de Melo Lisboa, H., A review of odour impact criteria in selected countries around the world. *Chemosphere*, **168**, pp. 1531–1570, 2017. DOI: 10.1016/j.chemosphere.2016.11.160.

[10] Zarra, T., Galang, M.G., Ballesteros, F., Belgiorno, V. & Naddeo, V., Environmental odour management by artificial neural network: A review. *Environ. Int.*, **133**, 105189, 2019. DOI: 10.1016/j.envint.2019.105189.

[11] Brancher, M., Piringer, M., Franco, D., Belli Filho, P., De Melo Lisboa, H. & Schauberger, G., Assessing the inter-annual variability of separation distances around odour sources to protect the residents from odour annoyance. *J. Environ. Sci. (China)*, **79**, pp. 11–24, 2019. DOI: 10.1016/j.jes.2018.09.018.

[12] Invernizzi, M., Brancher, M., Sironi, S., Capelli, L., Piringer, M. & Schauberger, G., Odour impact assessment by considering short-term ambient concentrations: A multi-model and two-site comparison. *Environ. Int.*, **144**, 105990, 2020. DOI: 10.1016/j.envint.2020.105990.

[13] Wojnarowska, M., Plichta, G., Sagan, A., Plichta, J., Stobiecka, J. & Sołtysik, M., Odour nuisance and urban residents' quality of life: A case study in Kraków's in Plaszow district. *Urban Clim.*, **34**, 100704, 2020. DOI: 10.1016/j.uclim.2020.100704.

[14] Schauberger, G. et al., Impact of global warming on the odour and ammonia emissions of livestock buildings used for fattening pigs. *Biosyst. Eng.*, **175**, pp. 106–114, 2018. DOI: 10.1016/j.biosystemseng.2018.09.001.

[15] Calafat, C. & Gallego-Salguero, A., Livestock odour dispersion and its implications for rural tourism: Case study of valencian community (Spain). *Spanish J. Agric. Res.*, **18**(2), pp. 1–17, 2020. DOI: 10.5424/sjar/2020182-15819.

[16] Vasile, A., Gheorghita, T., Babeanu, N. & Popa, O., Odour assessment for a pig farm through dynamic olfactometry and air dispersion modelling in order to reduce the odour pollution using biotechnologies. *Rev. Chim.*, **71**(1), pp. 61–66, 2020. DOI: 10.37358/RC.20.1.7812.

[17] Douglas Kreis, R., Limiting the environmental impact of animal production odors. *Environ. Int.*, 1(5), pp. 247–275, 1978. DOI: 10.1016/0160-4120(78)90026-0.

[18] Nicell, J.A., Assessment and regulation of odour impacts. *Atmos. Environ.*, **43**(1), pp. 196–206, 2009. DOI: 10.1016/j.atmosenv.2008.09.033.

[19] Murguía, W., Contaminación por olores: el nuevo reto ambiental. *Gac. ecológica*, **82**, pp. 49–53, 2007. http://www2.inecc.gob.mx/publicaciones/gacetas/521/olores.pdf.

[20] Ortega Vélez, A.-F., Análisis De La Inmisión De Olores Ofensivos En El Ramal 'A' Del Estero Salado De La Ciudad De Guayaquil Y Propuesta De Mitigación, 2016.

[21] Bokowa, A. et al., Summary and overview of the odour regulations worldwide. **12**(2), 2021.

[22] Ministerio de Ambiente & Desarrollo Sostenible, Resolución 2087 del 16 de diciembre de 2014, **53**(9), pp. 1689–1699, 2014.

[23] SurveyMonkey, SurveyMonkey sample size calculator, 2020.

[24] R Core Team, R: A language and environment for statistical computing. R Foundation for Statistical Computing, Vienna, 2020. https://www.r-project.org.

[25] Urrego Ortiz, E.L., Propuesta De Una Guia Teórica Y Práctica Para El Diseño De La Planificación En Un Sistema De Gestión Ambiental Para Granjas Porcicolas En El Departamento De Cundinamarca, 2016.

[26] PorkColombia, Guía de mejores técnicas disponibles para el sector porcícola en Colombia, 2015. https://www.porkcolombia.co/wp-content/uploads/2018/07/Guia-MTD-en-la-Porcicultura-de-Colombia.pdf.

LOCAL MONITORING OF TRAFFIC-RELATED AIR POLLUTION AROUND SCHOOLS IN SOUTH EAST LONDON, UK

HO YIN WICKSON CHEUNG & LIORA MALKI-EPSHTEIN
University College London, UK

ABSTRACT

Outdoor air quality (OAQ) presents a significant challenge for public health globally, especially in urban areas, with road traffic acting as the primary contributor to air pollution. Several studies have documented the antagonistic relation between traffic-related air pollution (TRAP) and the impact on health, especially to vulnerable members of the population, particularly young pupils. Generally, TRAP could restrict the ability of schoolchildren to learn and, more importantly, cause detrimental respiratory disease in their later life. But little is known about the specific exposure of children commuting to school and during the school day and the impact this may have on their overall exposure to pollution at a crucial time in their development. This project has set out to examine the air quality across primary schools in south east London (due to their massively increasing amount of redevelopment and population) and assesses the variability of data found based on their geographic location and surroundings. Nitrogen dioxide (NO_2) and PM contaminants ($PM_{2.5}$ and PM_{10}) were collected with diffusion tubes and portable monitoring equipment for eight schools across three local areas: Greenwich, Lewisham and Tower Hamlets. This study first examines the morphological features of the schools surrounding), then utilises two different methods to capture pollutant data. Moreover, comparing the obtained results with existing data from the London Air Quality Network (LAQN) to understand the differences in air quality pre- and post-pandemic. Most studies in this field have unfortunately neglected human exposure to pollutants and calculated referring to values from fixed monitoring stations. This paper introduces an alternative approach by calculating human exposure to air pollution from real-time data obtained when commuting around selected schools (driving routes and field walking).
Keywords: geographical feature, human exposure, schools, traffic related air pollution.

1 INTRODUCTION

The phenomenal increase in urbanisation in the past few decades, has consequently enlarged the population and the need for land transportation. However, ample evidence has pointed out that land transport is the main contributor to the elevated of air pollution in urban areas [1]. Pollution arising from the emissions of motor vehicles and road transportation (fossil fuel combustion) is typically referred to as traffic-related air pollution (TRAP) [2]. This type of air pollution can vastly increase mortality from stroke or heart disease and promote the development of cardiovascular diseases and respiratory illnesses, causing seven million people worldwide every year to die from this and 90% of the population to breathe air that exceeds the World Health Organization guideline limits containing high levels of pollutants [3].

Ten thousand pupils at more than 800 education institutions in London are exposed to extremely high levels of NO_2, well exceeding the EU legal limit of 40 $\mu g/m^3$ during school operating hours [4]. Among these institutions, over 400 primary schools with pupils under the age of 11 were designated, and 25% of schools were even found to be located in areas with dangerously high levels of air pollution [7]. Excessive inhalation of TRAP can cause a range of health issues. Short-term exposure can lead to the aggravation of existing respiratory problems. Long-term exposure could be linked to a greater susceptibility to infections of

WIT Transactions on Ecology and the Environment, Vol 259, © 2022 WIT Press
www.witpress.com, ISSN 1743-3541 (on-line)
doi:10.2495/AWP220071

respiratory disease [5]. A growing number of studies have shown an association between TRAP with the exacerbation of respiratory (asthma) and allergic symptoms in children [6]. Young pupils are at an exceptionally high risk of exposure to air pollution. As their organs (lungs) are still developing, they are more at risk [7].

Ambient concentrations of many air pollutants are elevated near roadways. This has particularly caught scientists' attention to school children living in urban areas, as a number of them are living near a congested roadway, and required to commute and study in a school near such road. Concentrations of pollutant contaminants that occur in microenvironments, such as inside street canyons or buildings or in-vehicle, have been shown to be higher than those measured at fixed-site monitors. Background concentration from such street monitoring stations might not accurately represent the actual distribution in the street, due to the sensitivity of airflow to local street geometry [8], therefore, localised monitoring is required. Although multiple studies have monitored personal exposure to air pollutants in vehicles and during commutes by foot, most of them focused on carbon monoxide, particulate matter and volatile organic compounds [9]–[11]. While the personal exposure to nitrogen dioxide remains largely unexplored. Periods for children commuting from home to schools often coincide with traffic congestion peaks, studies have found these would result in the commuters receiving a large proportion of their daily TRAP exposures, despite the short commuting time [11]. An exposure model conducted for in-vehicle commutes on a Los Angeles school bus route from Behrentz et al. [12] and Sabin et al. [13], has determined that the school bus commute contributed 10% of the total daily NO_2 exposure of schoolchildren, and 15% of total daily $PM_{2.5}$ exposures.

Air pollution in London has improved in recent years as a result of policies reducing emissions, nevertheless, the most recently updated Annual Pollution Map from London Air Quality Network [14] (Fig. 1(a)) has shown significant exceedances of the annual mean of NO_2 contaminants, on main roads and all around them. With south east London benefitting from regeneration/ redevelopment over the last decade, the population and the ongoing pressure on road traffic in the area have highly been increased. As such, south east London has been found to consist of the two most traffic-congested routes, the Blackwall Tunnel and the A2 routes. Mainly, due to the lack of routes selection for travelling to central London. Especially, the Blackwall Tunnel, where drivers travelling across are facing daily queues with average delays of 20 minutes.

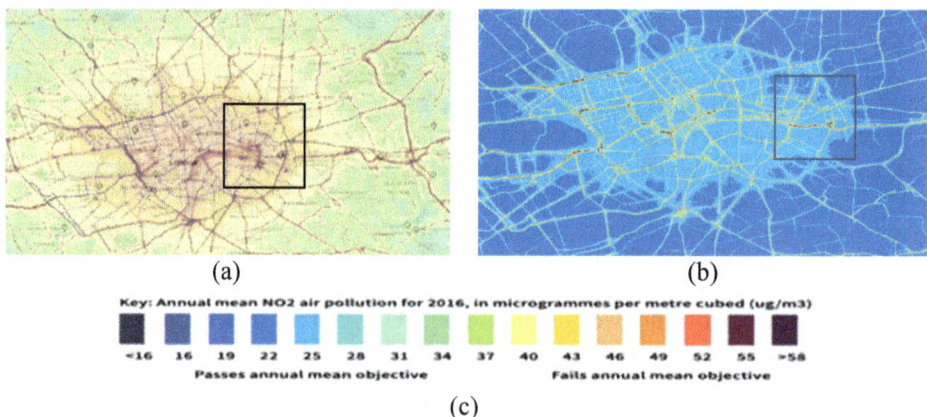

(a) (b)

Key: Annual mean NO2 air pollution for 2016, in microgrammes per metre cubed (ug/m3)

<16 16 19 22 25 28 31 34 37 40 43 46 49 52 55 >58

Passes annual mean objective Fails annual mean objective

(c)

Figure 1:   (a) Annual pollution map 2016; (b) Projected pollution map 2025; and (c) NO_2 index.

The SARS-CoV-2 (COVID-19) pandemic has ravaged the UK since late January 2020. The national lockdown alongside the public behaviours shifting towards work from home has significantly lowered vehicle traffic on roads in the UK. Motor vehicle usage nationally has reduced on average by 48%, meanwhile, the mean NO_2 concentrations have significantly reduced by 32% to 36% compared to the average of the previous seven years (2013–2019) [15]. This study aims to examine the localised exposure to TRAP around a selection of primary schools in south east London, in relation to the urban geometry, geographical features, and road configuration around these schools. The study was carried out in the context of a nationwide NO_2 reduction as a consequence of the COVID-19 pandemic, therefore this data is compared against the projected NO_2 annual map 2025 (2018; Fig. 1(b)), to highlight the benefits of a local reduction in emissions to reducing the actual exposures of schoolchildren.

2 METHODOLOGY

Nitrogen dioxide (NO_2) and PM contaminants ($PM_{2.5}$ and PM_{10}) were assessed in this study across eight schools in south east London. These schools are located in three main local areas across south east London: Greenwich, Lewisham and Tower Hamlets. The characteristics that schools are located in different areas allow them to be divided into batches when comparing results to understand how geographical factors can induce differences in pollution data. The overview of selected schools is shown in Fig. 2, with yellow placemarks representing their locations.

| (a) | (b) |

Figure 2: Overview for selection of schools. (a) Whole of London; and (b) SE London.

The selection of schools is presented in Table 1. These schools are selected based on research completed in 2017 commissioned by the Mayor of London [7]. This research has ranked 3,261 educational institutions (schools, nurseries, colleges and universities) in London (1 being the school with the cleanest air surrounded and 3,261 being the school suffering from the worst air pollution). Data from this research has indicated that 802 out of 3,261 educational institutions, which included 360 primary schools, were within 150 m of nitrogen dioxide pollution spots that have pollutant levels that exceed the EU annual legal limit of 40 $\mu g/m^3$. Primary schools ranked the worst across the three main local areas and located closest to the major congested road routes (A2, A12 and Blackwall Tunnel) were selected. In addition, this paper also describes an assessment of the NO_2 exposures against the EU guidelines, where hourly and annually standards are, respectively, 200 and 40 $\mu g/m^3$ [16]. From Table 1 it can be observed that the majority of these schools are in areas with NO_2 exceeding the above stated annual legal limits.

Table 1: Selection of schools.

School ID	Schools	Local Area	Mayor's Research (2017)	
			Ranking	NO_2 ($\mu m/m^3$)
1	Invicta Primary School	Greenwich	3035	47.3
2	Millennium Primary School	Greenwich	2443	39.9
3	St Mary Magdalene C of E School	Greenwich	2490	40.3
4	Kender Primary School	Lewisham	2928	45.1
5	St James Hatcham CE School	Lewisham	3069	48.1
6	Bow School	Tower Hamlet	3152	52.1
7	Marner Primary School	Tower Hamlet	3095	48.9
8	Woolmore Primary School	Tower Hamlet	3240	61.8

2.1 Data collection

2.1.1 Passive monitoring

Nitrogen dioxide diffusion tubes that monitored the concentrations of NO_2 continuously were distributed across the eight schools, with a total number of 10 diffusion tubes installed. A location study was completed on the surrounding area of each school to select the optimal installation point prior to the setup of diffusion tubes. The NO_2 sampling method was molecular diffusion, allowing compounds to move from an area of high concentration (air) to an area of low concentration (diffusion tubes). The absorbent used was 50% triethanolamine/acetone and the desorption efficiency (d) is 0.98. To allow proper molecular diffusion, the tubes were installed at least 2–3 m above ground level. Under the technical restrictions that sufficient concentration of NO_2 compounds is required to be absorbed onto the tubes for detection during the laboratory analysis, long-term monitoring (duration of 2–4 weeks) is necessary. As such, the 10 sets of tubes were set up from 29 June to 19 July 2021. An insight into the location of diffusion tubes installed near a typical congested road around the school is shown in Fig. 3. The date and time of installation and removal of the equipment are recorded to calculate the overall exposure of NO_2 in the installed time period. Additional ODA data was obtained from the UK Automatic Urban and Rural Network (AURN) for the assessment of pre- and post-pandemic air quality. Specifically, the John Harrison Way,

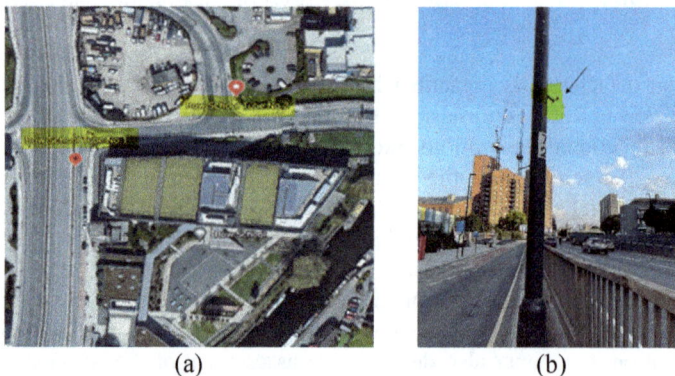

(a)   (b)

Figure 3:   (a) GPS location for diffusion tube location; and (b) Bow School diffusion tube number 1.

Woolwich Flyover, New Cross and Blackwall monitoring stations were selected across the three regional areas.

2.1.2 Active monitoring

TRAP was monitored both inside (commuting in vehicle) and outside vehicles (walking around the schools) during each route. Real-time data were collected for NO_2, PM_{10} (TSI environmental monitor EVM-7) and $PM_{2.5}$ (TSI aerosol monitors SidePak AM520). The logging intervals were 1 second and 5 seconds for AM520 and EVM-7 respectively, and were averaged over the duration of each route. The resolution for the toxic gas sensor (NO_2) is 0.1 PPM and the resolution of both devices' particulate sensors is 0.001 mg/m^3. Both these devices collect real-time aerosol concentration using a 90° optical light scattering photometer to determine the total mass concentration. Although the particulate matter was measured in mg/m^3 and NO_2 was measured in PPM, these values are presented in μg/m^3, as these units are commonly used in air quality guidelines. To avoid technical difficulties with the equipment, field monitoring was only undertaken during rain-free days and with background humidity under 95%. In-vehicle monitoring was conducted in a typical diesel saloon vehicle, with monitors positioned in the passenger seat. All windows were opened, meanwhile, both the recirculation and heating/cooling system were kept in the off position at all times.

2.2 Route selection

As minimal studies have focused their monitoring efforts specifically on characterised school operating hours, the field monitoring undertaken for this paper has taken place twice each day, during the morning (07:30–09:00) and afternoon school peak traffic hours (14:00–15:30). Three visits (experiment) were undertaken for PM monitoring and one was completed for NO_2 monitoring. Field monitoring was conducted on weekdays in August 2021 (summer). Dedicated routes were assigned for both driving and walking routes, driving routes 1 and 2 are shown in Fig. 4(a) and 4(c) respectively. Route 1 (A12) mainly focused on the Tower Hamlets area and the Blackwall Tunnel approach, whereas route 2 (A2) focused on the Lewisham area. In addition, route 1 was designed to investigate OAQ in highways (A102 and A12), and route 2 was designed to observe ordinary roadways. The sequential order of travelling for route 1 was school numbers 2, 3, 8, 6 and 7, whereas route 2 was school numbers 1, 5 and 4.

The driver stopped at each school along the designated route and commenced a 5–10 minute walk around the surroundings of the school. The monitors were turned on at the beginning of the drive and remained logging continuously until the end of the route. To minimise variability in results, walking paths were also assigned to follow for each school, an example of this for route 2 is shown in Fig. 4(b).

2.3 Morphological characteristics for the school surroundings

Analysis of the surrounding urban morphology was conducted with 3D images and tools from Google Earth. Major roads suffering from the most traffic congestion across each school were first identified. Then the distances between the school and these commuter roads were estimated along their vertical or horizontal direction. An example of such analysis for Bow School is shown in Fig. 5(a).

To understand the coverage of urban morphological features (e.g., road structure, green infrastructure, railway, and constructions sites) around each school, 3D images were

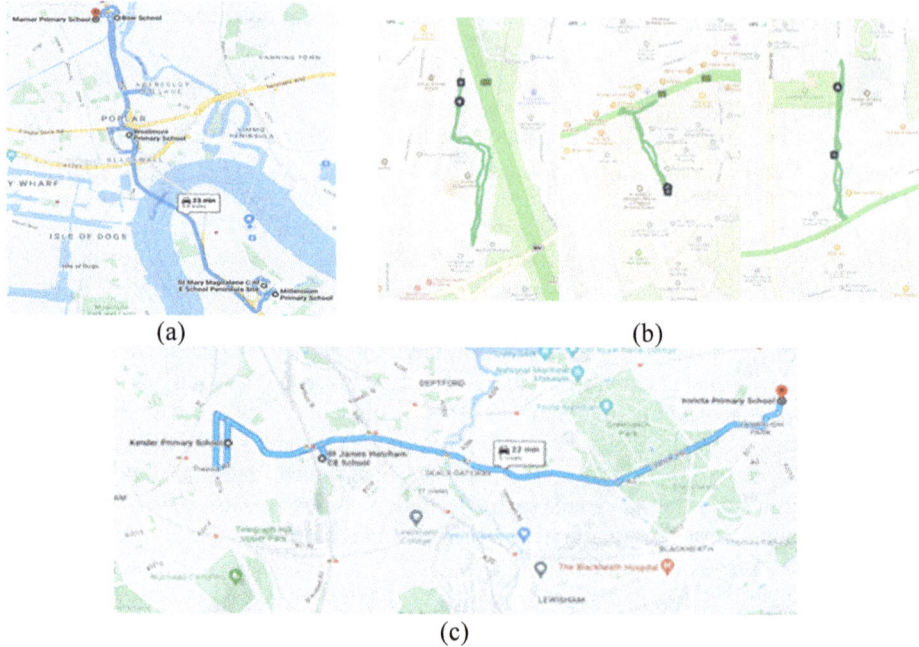

(a) (b)

(c)

Figure 4: (a) Route 1: Driving; (b) Route 2: Walking paths; and (c) Route 2: Driving.

(a) (b)

Figure 5: Local geometry of Bow School surrounding. (a) Distance to the main road; and (b) School surrounding shown in the 3D image (with urban road structure highlighted in blue).

procured from Google Earth at a scale of 1 to 40 m. The schools were placed in the middle of the image and the road structures surrounding the school were highlighted in light blue (Fig. 4(b)). The percentage of each feature compared to the total surface area of the images or school surroundings was conducted.

3 RESULT
Outdoor pollutants monitored in this paper were given colour indexes to be compared against the EU NO_2 and PM guidelines, where green, yellow and red indicate the pollutant concentration is safe, nearly exceeding and has exceeded the legal limit respectively.

3.1 Air quality: Fixed monitoring of nitrogen dioxide

Ten diffusion tubes were placed across the eight selected schools, where two diffusion tubes were installed around school numbers 4, 5 and 6. And one diffusion tube was installed around school numbers 2 and 3, as they are closely located next to each other. The majority of the schools have recorded a downwards trend for NO_2 concentrations when compared to the Mayor of London's research in 2017 (Table 1). This was expected, due to the public behaviours changes towards remote working. However, it can still be agreed that pollutants were still exceeding the EU legal guidelines. And even with the national lockdown restriction in place (during the period when diffusion tubes were installed), a significant increase of NO_2 reading can be observed in Bow School surroundings. Outdoor NO_2 readings from monitoring stations were obtained from the LAQN, chosen from the closest stations to the schools and are presented in Table 2. The John Harrison Way and Woolwich Flyover station are located in Greenwich near school numbers 1, 2 and 3. The New Cross station is located in Lewisham next to school numbers 4 and 5. The Blackwall station is located in Tower Hamlet close to school numbers 6, 7 and 8. Finally, NO_2 data in 2019 and 2021 are compared in terms of the pre- and post-COVID-19 pandemic air quality. The results show that all monitoring stations have recorded a reduction in NO_2 readings in 2021.

Table 2: Background and diffusion tubes NO_2 concentrations.

Schools	Monitors No.	NO_2 (µg/m³)	Station	2019 NO_2 (µg/m³)	2021 NO_2 (µg/m³)	UK Background Annual Mean (2020) (LAQN, 2022)	
				Outdoor Monitoring Stations Annual Mean			
1	1	33	Woolwich Flyover	52	39	Background	NO_2 (µg/m³)
2	2	35	John Harrison Way	33	22	Urban background	15
3							
4	3 (Minor Rd.)	19	New Cross	38	30	Rural background	5
	4	33					
5	5 (Minor Rd.)	22				Urban Traffic	23
	6	43					
6	7	62	Blackwall	47	38		
	8 (Minor Rd.)	28					
7	9	35					
8	10	39					

Note: (Minor Rd.) refers to diffusion tubes that were installed in a relatively less traffic-congested road (e.g., Bow School diffusion tube number 2 shown in Fig. 5(a))

3.2 Human exposure: Mobile monitoring

To calculate the total exposure during the entire commuting period, for each specific segment of travel, the average values of NO_2 pollutants were multiplied by the total time spent by an individual (eqn (1)). Where ε is integrated exposure over the specific period, C_j equals the concentration experienced in environment j, t_j is the time (minutes) in environment j, and J represents the total number of environments occupied over a specific time. For example, the average recorded during "Walk 1" (A) is multiplied by the time (T) taken during this associated action. This step (A × T) is repeated for every following action. The exposures obtained are summed up to indicate the total exposures for the whole journey.

$$\varepsilon = C_j t_j. \tag{1}$$

Examples of the exposure throughout the journey with detailed timesteps are shown in Fig. 6. The actual time taken for commuting in-vehicle between schools (drive) and walking around each school (walk) were recorded and these are presented in Table 3. The hourly exposures were obtained by dividing the exposure of each associated action by the time (T) and multiplied by 60 minutes. This allows the exposures to be presented in hourly

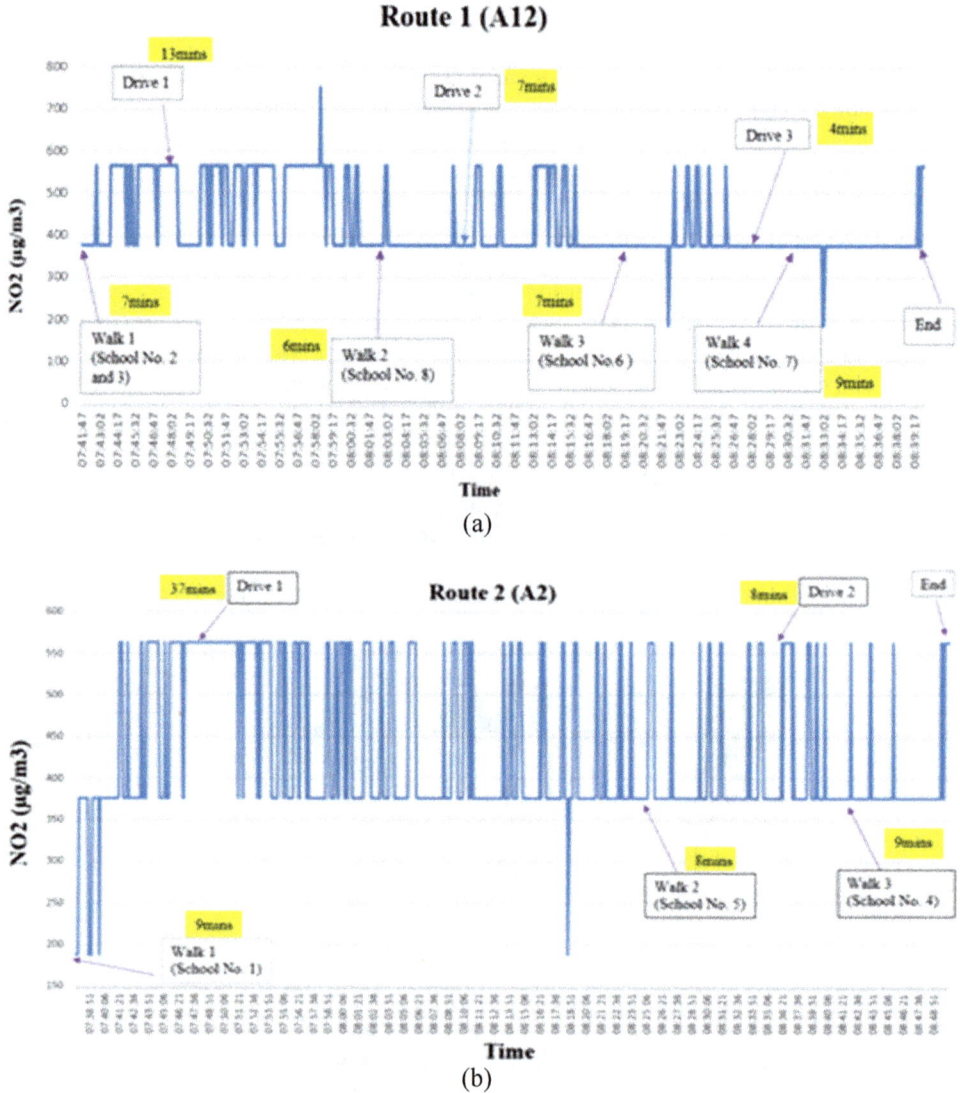

Figure 6:   Example of NO$_2$ exposure in detailed time steps. (a) Route 1; and (b) Route 2.

averages, to compare against the EU NO$_2$ hourly standard of 200 μm/m^3. The variability between morning and afternoon data is expected to be caused by the difference in traffic volume, leading to a longer commuting time.

3.3 Air quality: Particulate matter concentrations

The field experiment was undertaken three times to measure the PM concentrations during commuting to school and to evaluate against the EU Particulate Matter Guideline, where the annual guidelines for PM$_{2.5}$ and PM$_{10}$ are, 25 and 40 μg/m^3, respectively [16]. The PM$_{2.5}$ and

Table 3: Total NO$_2$ exposure for the entire route.

	Route 1 (A12)					
	Morning			Afternoon		
Action	Time (Mins)	Exposure (µg/m^3)	Mean Exposure/ Hr	Time (Mins)	Exposure (µg/m^3)	Mean Exposure/ Hr
Walk 1	7	3400	486	10	3100	310
Drive 1	13	6500	500	16	6800	425
Walk 2	6	2300	383	10	4000	400
Drive 2	7	2900	417	10	4200	420
Walk 3	7	2800	400	11	3700	333
Drive 3	4	1600	400	5	2400	480
Walk 4	9	3400	383	12	4300	358
	Total Exposure	23,000		Total Exposure	28,000	
	Route 2 (A2)					
Walk 1	9	3900	433	7	2000	283
Drive 1	37	17000	467	18	6100	333
Walk 2	8	3200	400	8	2400	300
Drive 2	8	3400	433	9	3100	350
Walk 3	9	3600	400	9	2800	317
	Total Exposure	31,000		Total Exposure	17,000	

Table 4: Outdoor average PM$_{2.5}$ and PM$_{10}$ (µg/m^3).

		1st Visit		2nd Visit		3rd Visit		Average PM Values	
		A.M.	P.M.	A.M.	P.M.	A.M.	P.M.	A.M	P.M
Route 1	PM 2.5	47	28	22	51	43	35	37	38
	PM 10	10	80	40	10	50	30	33	40
Route 2	PM 2.5	57	38	29	36	44	45	43	40
	PM 10	20	20	20	30	10	90	17	47

Note: A.M. refers to morning measurements; and P.M. refers to afternoon measurements.

PM$_{10}$ across each route are presented as averages in Table 4. Results show that values for PM$_{10}$ were generally under the annual guideline, whereas readings for PM$_{2.5}$ tend to significantly go above the EU standards.

3.4 Distance to closest major road and other urban morphological features

The urban morphology characteristics of each school are presented in Table 5. The two closest major roads were identified according to their average daily traffic volume. And the distance between each road and the schools were measured. All schools were found to be within 300 m of a major road, with school number 7 as close as 30 m. However, a number of these roads were highways that have larger traffic volumes (e.g., A12 has an approximate daily traffic flow of 80,000 vehicles) and contribute to higher emissions. The influence of railway services on the concentration of toxic pollutants to school pupils should not be

Table 5: Geographical characteristics of school surrounding.

	Distance (meter) to the closest major road to school				Percentage (%) coverage of geographical features				Mean NO$_2$ (µg/m^3)		Mean PM (µg/m3)	
School	1st Road	Distance	2nd Road	Distance	Road	Green	Other	Notes	Monitors	Stations	PM10	PM2.5
1	B210	215	A102	40	18	40	-	-	33	39	32	42
2	Bugsby's Way	110	A102	245	19	20	-	-	35	22	37	38
3	Millennium Way	110	A102	245	11	10	35	Construction Site	35	22	37	38
4	A202	210	A2	210	13	30	5	Construction Site	26	30	32	42
5	Railway	60	A2	130	27	25	5	Railway	33	30	37	42
6	A12	55	-	-	20	20	10	Parking Slot	45	38	37	38
7	A12	130	Devas Street	30	16	15	15	Parking Slot	35	38	37	38
8	A102	55	-	-	55	20	10	Parking Slot	39	38	37	38

Note: "Notes" provide additional detail to the features that "Others" are indicating; Grey highlight indicates that the associated road is considered a highway.

disregarded, as such, the distance of school number 5 to the nearby railway track was also measured.

Green infrastructure could improve air quality by providing barriers to sources of pollution. Therefore, to determine the variability of air quality in relation to geographical features, an estimation of the percentage of coverage to road structure, green infrastructure and other possible contributors to TRAP are conducted. Construction sites, parking slots and railways were identified to be other possible factors that could affect the quality of air around schools and therefore were defined as other (in Table 5). From the results, school number 8 has the highest percentage of road structure in its surroundings. Additionally, results presented have illustrated that NO_2 concentrations are much higher in schools surrounded by highways (school numbers 2, 3, 6, 7, 8), whereas $PM_{2.5}$ concentrations are often elevated in schools with ordinary roadways surrounded (school numbers 1, 4, 5).

4 DISCUSSION

The NO_2 concentrations monitored with diffusion tubes were used to assess the outdoor air quality at each school, by comparing data against values from fixed-site monitoring stations (Table 2). All of the schools have shown higher concentrations locally when compared to the closest available station, except for schools 1 and 7, this is likely due to the installed diffusion tubes for schools 1 and 7 being located at a side road (Table 2), which was far apart from where the enormous amount of traffic was congested and against the direction of dispersion of air pollutants. Furthermore, Table 2 indicated that the UK NO_2 average across the three different backgrounds was 14.3 $\mu m/m^3$, whereas the mean of data from local monitoring stations was 32.25 $\mu m/m^3$, which is a significant increase of 126%. Whilst the NO_2 average of localised monitors across the selection of schools was 34.9 $\mu m/m^3$ (an 8% increase when compared to the average of local monitoring stations). Hence, the significant increase in the averages from localised micro-environment measurements have consolidated the initial assumptions to the importance of localised monitoring.

The individual human exposures during commuting for routes 1 and 2 are shown with detailed timesteps in Fig. 6, and these demonstrate that NO_2 measurements are observed to be generally at their peak during commuting in a vehicle. The mean exposure per hour when travelling in routes 1 and 2 is found to be at least 55% and 41.5% above the EU NO_2 hourly legal limit of 200 $\mu m/m^3$, respectively.

The measured $PM_{2.5}$ concentrations for both routes (Table 4) were up to an order of magnitude higher than the suggested guidance value of 25 $\mu m/m^3$, whereas only some PM_{10} concentrations were above the legal limit. This has suggested that TRAP is more likely to result in an elevated $PM_{2.5}$ concentration.

It is found from the local microenvironment measurements and the urban morphological analysis that schools located highly close to highways (route 1) might not necessarily experience the worst air pollutants. Instead, the worst traffic congestion nearby (route 2) might better explain poor air quality. For example, 37 minutes were taken to commute from school numbers 1–5 during morning commuting for route 2 (Fig. 6 and Table 3). The longer travelling time has resulted in a total exposure of 31,000 $\mu m/m^3$, which is almost twice the NO_2 exposure of afternoon commuting in route 2 and is 35% higher than that of route 1 morning commuting, suggesting the negative correlation of traffic volume and congestion to air quality.

Route 1 was designed to investigate OAQ in highways (Blackwall Tunnel (A102 and A12)), and route 2 was designed to observe ordinary roadways. The results indicated that, although route 1 is one of the busiest and most congested highways in London, no significant difference can be observed between the average $PM_{2.5}$ values recorded on route 1 when compared with route 2 (Table 4). This is likely due to the narrow street canyon geometry across streets near the A2, but further research is necessary to understand the causes.

As shown in Table 5, the NO_2 concentrations were higher in schools having major highways nearby (school numbers 2, 3, 6, 7 and 8). In contrast, PM values were relatively higher for the other schools. Moreover, the school with the surrounding covered by the highest percentage of road structure did not have elevated pollutant concentrations. Similarly, schools surrounded by a larger proportion of vegetation did not result in a lower pollutant concentration. These have suggested that the effect of coverage of localised road structure and green infrastructure to OAQ may be limited, but the sample size was too small to evaluate this with certainty.

5 CONCLUSION

In general, our findings suggested that school pupils commuting to schools may be repeatedly exposed to elevated levels of traffic-related air pollutants, especially to the exceedance of $PM_{2.5}$ and the hourly exposures of NO_2. Furthermore, suggesting that traffic characteristics, land usage and morphology are some of the important determinants to the elevated of TRAP exposures. To conclude, this paper has yielded significant benefits to understanding air quality across schools in south east London by determining the exposure for schoolchildren during their daily commuting. The importance of this work is to confirm the severity of TRAP relating to schoolchildren and promote the necessity of considering environmental sustainability for policymakers during decision making.

It is recommended that a statistically significant set of schools be assessed for their local urban morphology, wind direction and speed, and traffic volume, to determine whether there is a clear link to the effect of geographical features on OAQ.

ACKNOWLEDGEMENT

This work was supported by activities of the TAPAS air quality network, NERC reference NE/V002341/1.

REFERENCES

[1]   Chin, A.T., Containing air pollution and traffic congestion: Transport policy and the environment in Singapore. *Atmospheric Environment*, **30**(5), pp. 787–801, 1996.
[2]   National Toxicology Program, Traffic-related air pollution and hypertensive disorders of pregnancy, 2021. https://ntp.niehs.nih.gov/whatwestudy/assessments/noncancer/completed/pollution/index.html. Accessed on: 17 Aug. 2021.
[3]   World Health Organization, Air pollution, 2021. https://www.who.int/health-topics/air-pollution#tab=tab_1. Accessed on: 27 Jul. 2021.
[4]   Taylor, M. & Laville, S., Revealed: Thousands of children at London schools breathe toxic air, 2017. https://www.theguardian.com/uk-news/2017/feb/24/revealed-thousands-of-children-toxic-air-london-nitrogen-dioxide. Accessed on: 10 Mar. 2022.
[5]   Brown, L., Barnes, J. & Hayes, E., Traffic-related air pollution reduction at UK schools during the COVID-19 lockdown. *Science of The Total Environment*, **780**, 146651, 2021.
[6]   Wood, H., Marlin, N., Mudway, I., Bremner, S., Cross, L., Dundas, I., Grieve, A., Grigg, J., Jamaludin, J., Kelly, F., Lee, T., Sheikh, A., Walton, R. & Griffiths, C., Effects of air pollution and the introduction of the London low emission zone on the prevalence of respiratory and allergic symptoms in schoolchildren in east London: A sequential cross-sectional study. *PLoS One*, **10**(8), 0109121, 2015.

[7]   Mayor of London, The Mayor's school air quality audit programme, 2018. https://www.london.gov.uk/sites/default/files/180521_saq_report_-_marner_primary_tower_hamlets_appendicies_and_cover_updated.pdf. Accessed on: 4 Aug. 2021.

[8]   Karra, S., Malki-Epshtein, L. & Neophytou, M.K.A., Air flow and pollution in a real, heterogeneous urban street canyon: A field and laboratory study. *Atmospheric Environment*, **165**, pp. 370–384, 2017.

[9]   Marshall, J.D. & Behrentz, E., Vehicle self-pollution intake fraction: Children's exposure to school bus emissions. *Environmental Science and Technology*, **39**(8), pp. 2559–2563, 2005.

[10] Abi Esber, L., El-Fadel, M., Nuwayhid, I. & Saliba, N., The effect of different ventilation modes on in-vehicle carbon monoxide exposure. *Atmospheric Environment*, **41**(17), pp. 3644–3657, 2007.

[11] Alvarez-Pedrerol, M., Rivas, I., López-Vicente, M., Suades-González, E., Donaire-Gonzalez, D., Cirach, M., de Castro, M., Esnaola, M., Basagaña, X., Dadvand, P. & Nieuwenhuijsen, M., Impact of commuting exposure to traffic-related air pollution on cognitive development in children walking to school. *Environmental Pollution*, **231**, pp. 837–844, 2017.

[12] Behrentz, E., Sabin, L.D., Winer, A.M., Fitz, D.R., Pankratz, D.V., Colome, S.D. & Fruin, S.A., Relative importance of school bus-related microenvironments to children's pollutant exposure. *Journal of the Air and Waste Management Association*, **55**(10), pp. 1418–1430, 2005.

[13] Sabin, L., Behrentz, E., Winer, A., Jeong, S., Fitz, D., Pankratz, D., Colome, S. & Fruin, S., Characterizing the range of children's air pollutant exposure during school bus commutes. *Journal of Exposure Science and Environmental Epidemiology*, **15**(5), pp. 377–387, 2004.

[14] London Air Quality Network, Annual Pollution Maps, 2016. https://www.londonair.org.uk/LondonAir/Default.aspx. Accessed on: 16 Mar. 2022.

[15] Higham, J.E., Ramírez, C.A., Green, M.A. & Morse, A.P., UK COVID-19 lockdown: 100 days of air pollution reduction? *Air Quality, Atmosphere and Health*, **14**(3), pp. 325–332, 2021.

[16] European Environment Agency, Air quality standards, 2022. https://www.eea.europa.eu/themes/air/air-quality-concentrations/air-quality-standards. Accessed on: 16 Mar. 2022.

EVALUATION OF THE ECOLOGICAL STATE USING THE WATER QUALITY INDEX AND FLUVIAL HABITAT INDEX OF THE URBAN BASINS OF PANAMA

QUIRIATJARYN M. ORTEGA-SAMANIEGO[1,2], ANDRES FRAIZ[3], ARTURO DOMINICI[4],
HAYDEE OSORIO[5], ADRIAN RAMOS-MERCHANTE[6], EDGAR ARAUZ[7],
MARIA PACHES[1] & INMACULADA ROMERO[1]
[1]Research Institute of Water and Environmental Engineering, (IIAMA),
Universitat Politècnica de València, Spain
[2]Ministerio de Ambiente de Panamá, Panamá
[3]Wetlands International, Panamá
[4]Universidad Marítima Internacional de Panamá, Panamá
[5]Universidad Tecnológica de Panamá, Panamá
[6]University of Huelva, Spain
[7]Universidad de Panamá, Panamá

ABSTRACT

The evaluation of the ecological status of aquatic ecosystems through indices allows quantifying and assessing the environmental impacts caused by human activities, in order to create a baseline for future adjustments in policies regarding environmental management, both at the level of local governments as nationals, who can collaborate in a better management of water resources. This study was carried out in the urban basins of Panama City, a place with the highest population concentration in the country, specifically in the Matasnillo, Juan Díaz and Pacora Rivers, in order to assess the ecological status using the water quality index (WQI) and the river habitat index (FHI). With the data collected from the samples in the high, medium and low zones, a Kruskall–Wallis Test was applied to determine if there were significant differences between the sampling points and in the season's characteristics of the tropical climate, the humid and dry, plus a sampling in and transition between both seasons. The results show us that the rivers maintain better conditions in remote areas or with less human impact, indicating that the upper areas of the Matasnillo, Juan Díaz and Pacora Rivers have the best scores in the WQI and FHI, decreasing their score in the middle and lower zones. The low WQI and FHI scores indicate the environmental impact caused by alterations to habitats due to canalization or construction of civil structures that end up modifying the hydrology, reducing the heterogeneity of habitats and the type of substrate, in addition to inadequate waste management, solids and liquids from human settlements, commercial and industrial activities, as well as agricultural areas in the rural perimeter that as a whole exert pressure on these urban basins as they pass.

Keywords: ecological estate, water quality index (WQI), fluvial habitat index (FHI), spatio-temporal analysis, Matasnillo River, Juan Diaz River, Pacora River, Panama.

1 INTRODUCTION

The health and quality of the waters in rivers and streams are of great importance for both human well-being and nature, for which the implementation of multimeric models for ecosystem studies using biological and chemical variables have been used frequently, obtaining results favorable for the management and identification of the conditions in the courses of superficial waters [1]. There is no doubt that one of the great challenges of this century is to seek to maintain the natural biological structure and functional attributes of aquatic ecosystems [2], case that is increasingly difficult, considering the destruction of natural habitats attenuated by the presence of pollutants in rivers that cross agricultural, industrial and urban areas [3].

On the other hand, hydrological variation can play an important role in the structure and relationships of river organisms between hydrology and biology that are influenced by flow

WIT Transactions on Ecology and the Environment, Vol 259, © 2022 WIT Press
www.witpress.com, ISSN 1743-3541 (on-line)
doi:10.2495/AWP220081

regimes [4]. Being that according to Magoulick et al. [5], the hydrology–biology relationships must be examined within the flow regime. Furthermore, the variation of biological and hydrological characteristics at multiple spatial scales are important to conserve the natural flows of ecological relationships [6].

2 MATERIALS AND METHODS

2.1 Study area

2.1.1 Basin 142: Matasnillo River
The basin delimited between the Caimito and Juan Diaz Rivers, its extension is 383 km^2, the average elevation is 67 m above sea level and the highest point is 507 m above sea level, it is located in the Pacific of the Province of Panama. South main river is the Matasnillo that its total length is 6 km. It registers an average annual rainfall of 2,122 mm, it has a temperate tropical savannah climate, and its vegetation is made up of mature, secondary and little intervened forest); as well as grasslands and wetlands. This basin has other rivers of great importance such as the Curundu, Rio Abajo, Matias Hernandez and Cárdenas [7].

2.1.2 Basin 144: Juan Diaz River
The basin has an area of 322 km^2, located on the Pacific slope in the province of Panama. Its main river is the Juan Diaz, whose length is 22.5 km, with an average flow of 5.7 m^3 and average annual rainfall of 2,466 mm. The average elevation of 90 m above sea level and the highest point (800 m above sea level) is located between Cerro Azul and Cerro Jefe, it has a temperate tropical savannah and humid tropical climate. Among its tributaries are Las Lajas, María Prieta, Naranjal, Palomo, Quebrada Espavé and Malagueto [7].

2.1.3 Basin 146: Pacora River
It is located on the Pacific slope, within the province of Panama, in the district of Panama passing through the districts of Pacora and San Martín, between coordinates 8°00' and 8°20' north latitude and 79°15' and 79°30' west longitude. It maintains an area of 388 km^2. Among its natural limits we can mention: To the north, with the Chagres River basin; to the south, with the Bay of Panama; to the east, with the Bayano River basin; and to the west, with the Juan Díaz River basin. It is a basin considered a priority for the country, considering that it is the only source of surface water that supplies the Pacora water treatment plant. Its main course is the Pacora River, with a total length of 48 km, which rises in the mountains of the Cordillera Central to the Pacific Ocean, meeting its tributaries on its way, such as the Tataré River, Utivé River, Calobré River and Indio River, covering an area of 364 km^2. The highest point is Cerro Jefe (1,007 m above sea level). In the course of the basin there are three life zones: very humid premontane forest, humid premontane forest and tropical humid forest without the presence of protected areas [7].

2.2 Data analysis

2.2.1 Water quality index (WQI)

2.2.1.1 Spatial analysis of WQI
The data were collected in the wet season: September 2020, dry: March and transition: April 2021, in the high, middle and low areas of the basins 142: Matasnillo River, 144: Juan Diaz River and 146: Pacora River.

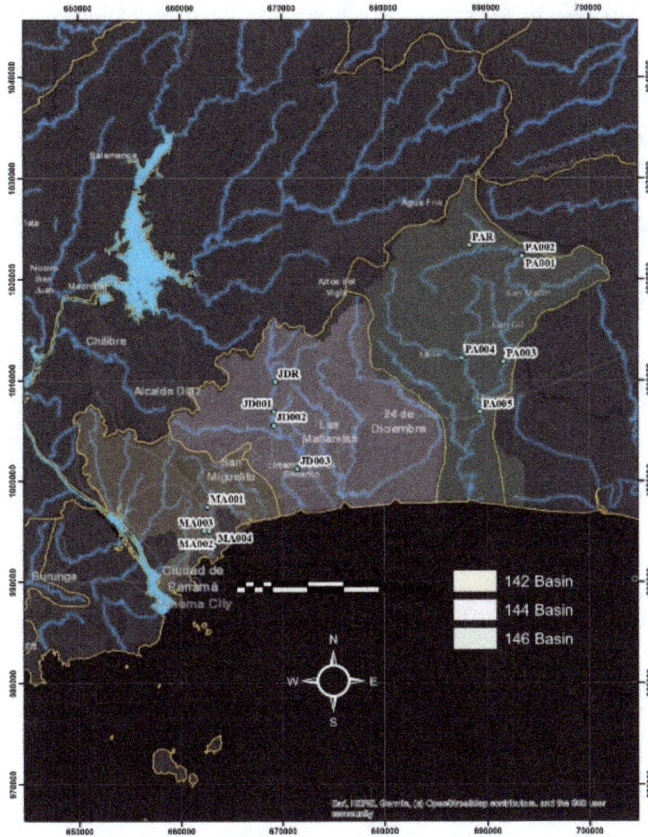

Figure 1: Location of the study sites.

The parameters used for the calculation of the WQI were T (°C), pH, DO (mg/L), fecal coliforms (NPM/100 mL), BOD_5 (mg/L), NO_3 (mg/L), and turbidity (NTU), using the standard methods as a guide [8]. Once the calculation is done, it will be categorized as highly polluted, polluted, slightly polluted, acceptable, and uncontaminated water quality.

The Kruskal–Wallis Test was applied to determine if there are significant differences for the WQI between points MA001, MA002 and MA003 Matasnillo River, JD001, JD002 and JD003 for Juan Diaz River and PA001, PA002, PA003, PA004, PA005 for Pacora River, using the statistical package PAST 4.01 and SPSS Version 28.

2.2.1.2 Temporal analysis of WQI

The Kruskal–Wallis Test was applied to determine if there are significant differences between the dry, wet and transition season of the WQI, using the statistical package PAST 4.01 and SPSS Version 28.

2.2.2 Fluvial habitat index (FHI)

The data were collected in the wet season: September 2020, dry: March and transition: April 2021, in the high, middle and low areas of the basins 142: Matasnillo River, 144: Rìo Juan Diaz and 146: Pacora River.

For the FHI curriculum, an adaptation was made to the methodology used by Pardo et al. [10], categorizing from 0–100, very good, good and does not reach very good.

2.2.2.1 Spatial analysis of FHI
The Kruskal–Wallis Test was applied to determine if there are significant differences for the FHI between points MA001, MA002 and MA003 in the Matasnillo River, JD0001, JD002 and JD003 in the Juan Diaz River, PA001,PA002, PA003, PA004, PA005 in the Pacora River, using the statistical package PAST 4.01 and SPSS Version 28.

2.2.2.2 Temporal analysis of FHI
The Kruskal–Wallis Test was applied to determine if there are significant differences between the dry, wet and transition season of FHI, using the statistical package PAST 4.01 and SPSS Version 28.

3 RESULTS AND DISCUSSION

3.1 Basin 142: Matasnillo River

3.1.1 Water quality index (WQI)
The WQI values are presented in Table 1, which shows that for the points MA001WE, MA001DR and MA001TR they present values of 60 to 69 corresponding to the slightly polluted ratings. For points M002WE, MA002DR, MA002TR, MA003WE, MA003DR, MA003TR, MA004WE, MA004DR, MA004TR they present values of 35 to 47 corresponding to the polluted ratings.

3.1.1.1 Spatio-temporal analysis of the WQI

3.1.1.1.1 Spatial analysis
There are significant differences between points MA001, MA002, MA003 and MA004 according to the Kruskal–Wallis Test as shown in Fig. 2. The minimum WQI values were presented at point MA004 and the maximum values at point MA001 as described in Table 2.

3.1.1.1.2 Temporal analysis of the WQI
The temporal variations are in Table 2, according to the Kruskal–Wallis Test there are no significant differences between the dry, wet and transitional seasons.

3.1.2 Fluvial habitat index (HFI)
The FHI values are presented in Table 1, which shows that for the points MA001WE, MA001DR and MA001TR they present values of 50 to 58 corresponding to the good ratings. For points M002WE, MA002DR, MA002TR, MA003WE, MA003DR, MA003TR, MA004WE, MA004DR, MA004TR they present values of 25 to 48 corresponding to the ratings of not reaching good.

Table 1: WQI and FHI ratings of the monitored sites.

COD	WQI	FHI	Site
MA001WE	60	58	
MA001DR	62	50	
MA001TR	69	58	
MA002WE	35	48	
MA002DR	38	48	
MA002TR	39	48	
MA003WE	41	33	
MA003DR	45	33	
MA003TR	42	33	
MA004WE	36	25	
MA004DR	47	33	
MA004TR	44	25	
JDR	93	98	
JD001WE	83	88	
JD001DR	89	86	
JD001TR	93	88	
JD002WE	72	65	
JD002DR	82	65	
JD002TR	79	60	

Table 1: Continued.

COD	WQI	FHI	Site
JD003WE	46	45	
JD003DR	70	45	
JD003TR	58	40	
PAR	91	86	
PA001WE	76	72	
PA001DR	83	72	
PA001TR	92	72	
PA002WE	74	58	
PA002DR	75	58	
PA002TR	75	58	
PA003WE	73	54	
PA003DR	77	54	
PA003TR	73	67	
PA004WE	73	61	
PA004DR	79	61	
PA004TR	77	54	
PA005WE	68	50	
PA005DR	78	50	
PA005TR	76	57	

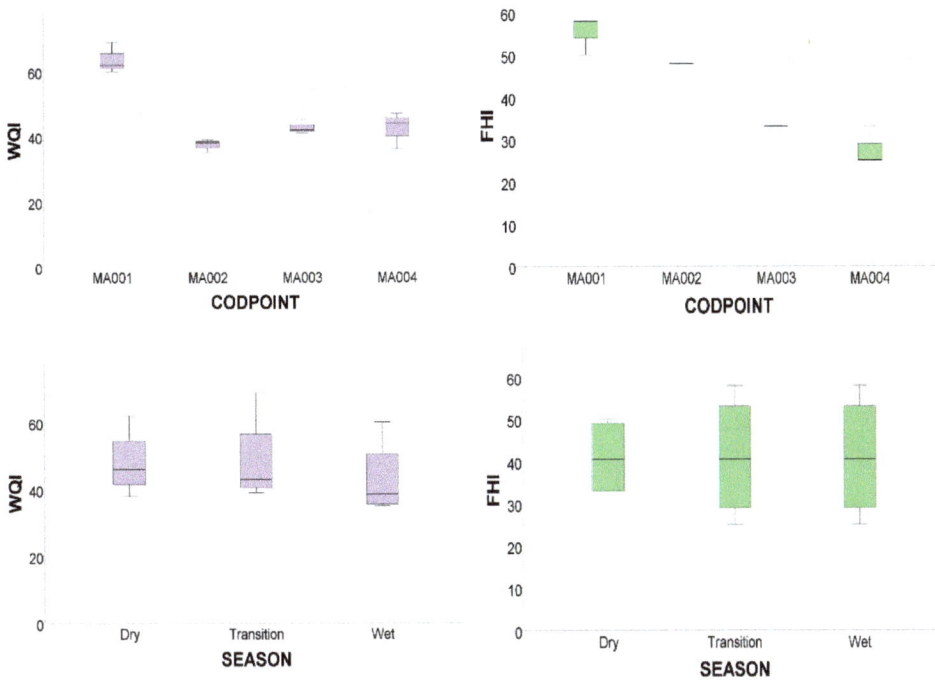

Figure 2: Boxplot of spatial-temporal variations of WQI and the FHI Matasnillo River.

Table 2: Descriptive statistics and Kruskal–Wallis of WQI and the FHI of Matasnillo River.

Index	N	Min.	Max.	Mean	Std. error	Variance	Std. dev.	Median	SEASON p (same)	CODPOINT p (same)
WQI	12	35	69	46.5	3.208653	123.5455	11.1151	43	0.4649	0.04344

3.1.2.1 Spatio-temporal analysis of FHI

3.1.2.1.1 Spatial analysis
There are significant differences between points MA001, MA002, MA003 and MA004 according to the Kruskal–Wallis Test as shown in Fig. 2.The minimum FHI values were presented at point MA004 and the maximum values at point MA001 as described in Table 2.

3.1.2.1.2 Temporal analysis
The temporal variations are presented in Fig. 2, according to the Kruskal–Wallis Test there are no significant differences between the dry, wet and transitional seasons.

According to the environmental monitoring carried out by ANAM [8], the Matasnillo River is classified as polluted to highly polluted, this due to domestic discharges and industrial activities, these waters being not suitable for use and aquatic life, except in the upper parts with restricted permission.

3.2 Basin 144: Juan Diaz River

3.2.1 Water quality index (WQI)

The WQI values are presented in Table 1, the JDR point is a reference point, which exemplifies the ideal conditions, and we have not considered it for the present analyses. As seen in the table for points JD001WE, JD001DR, JD002WE, JD002DR, JD002TR, JD003DR present values from 70 to 89 corresponding to the ratings of acceptable, for point JD001TR with a value of 93, Uncontaminated, JD003TR presents the value of 58 slightly polluted, point JD003WE its value is 46 with contaminated rating.

3.2.1.1 Spatio-temporal analysis of the WQI

3.2.1.1.1 Spatial analysis

There are significant differences between points JD001, JD002, JD003 according to the Kruskal–Wallis Test as shown in Fig. 3.The minimum WQI values were presented at point JD003 and the maximum values at point JD001 as described in Table 3.

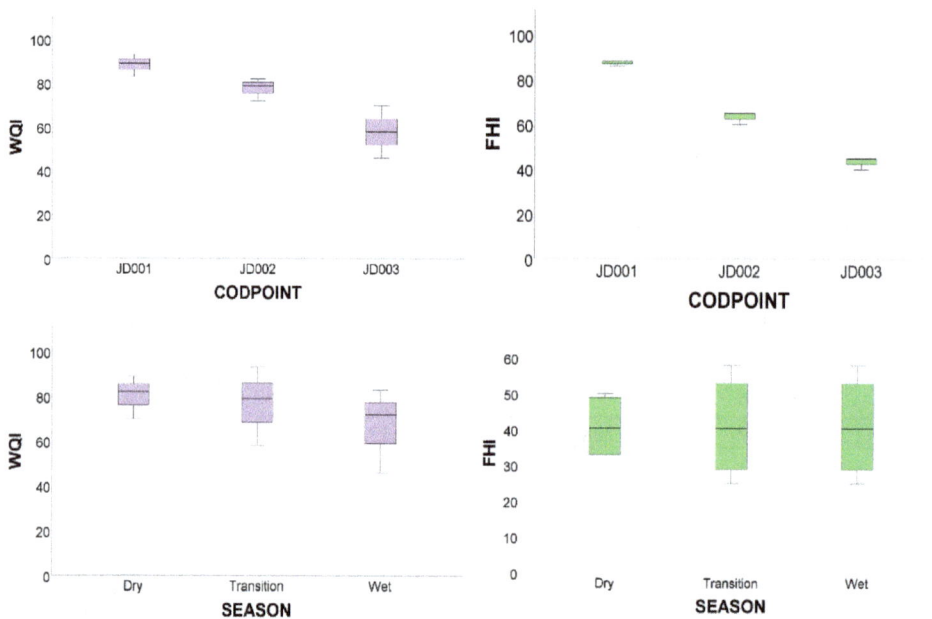

Figure 3: Boxplot of spatial-temporal variations of WQI and the FHI of Juan Diaz River.

Table 3: Descriptive statistics and Kruskal–Wallis of the WQI and FHI of Juan Diaz River.

Index	N	Min.	Max.	Mean	Std. error	Variance	Std. dev.	Median	SEASON p (same)	CODPOINT p (same)
WQI	9	46	93	74.66667	5.016639	226.5	15.04992	79	0.7326	0.02732
FHI	9	40	88	64.66667	6.398785	368.5	19.19635	65	0.9025	0.02491

3.2.1.1.2 Temporal analysis
The temporal variations are presented in Fig. 3, according to the Kruskal–Wallis Test there are no significant differences in WQI values between the dry, wet and transition seasons.

3.2.2 Fluvial habitat index (FHI)
The FHI values are presented in Table 1, which shows that for the points JD001WE, JD001DR and JD001TR they present values from 86 to 88 corresponding to the very good ratings. For points JD002WE, JD002DR, JD002TR have values of 60 to 65 corresponding to the ratings of good
    JD003WE, JD003DR, MA003TR present values of 40 to 45 corresponding to the ratings of not reaching good.

3.2.2.1 Spatio-temporal analysis of FHI

3.2.2.1.1 Spatial analysis
There are significant differences between points JD001, JD002, JD003 according to the Kruskal–Wallis Test as shown in Fig. 3. The minimum FHI values were presented at point JD003 and the maximum values at point JD001 as described in Table 3.

3.2.2.1.2 Temporal analysis
The temporal variations are presented in Fig. 3, according to the Kruskal–Wallis Test there are no significant differences between the dry, wet and transitional seasons.
    According to the analyses carried out by ANAM [8], they qualify as acceptable in the high areas, however the deterioration increases for the middle and lower areas of the river, presented ratings of contaminated or little contaminated, this due to the contributions of wastewater from population settlements, shops and industries.

3.3 Basin 146: Pacora River

3.3.1 Water quality index (WQI)
The WQI values are presented in Table 1, the PAR point with a rating of 93 is a reference, exemplifies the ideal conditions, and we have not considered it for the present analyses. As shown in Table 1 for the point PA001TR has a non-contaminated rating, points PA001WE, PA001DR, PA002WE, PA002DR, PA002TR, PA003WE, PA003DR, PA003TR, PA004WE, PA004DR, PA004TR, PA005DR, PA005TR have values of 73 to 83 corresponding to the ratings of acceptable, for point PA005WE with a value of 68 rating of contaminated.

3.3.1.1 Spatio-temporal analysis of WQI

3.3.1.1.1 Spatial analysis
There are no significant differences between points PA001, PA002, PA003, PA004, PA005 according to the Kruskal–Wallis Test as shown in Fig. 4.
    The minimum WQI values were presented at point PA005 and the maximum values at point PA001 as described in Table 4.

3.3.1.1.2 Temporal analysis
The temporal variations are presented in Fig. 4, according to the Kruskal–Wallis Test there are significant differences in WQI values between the dry, wet and transition seasons.

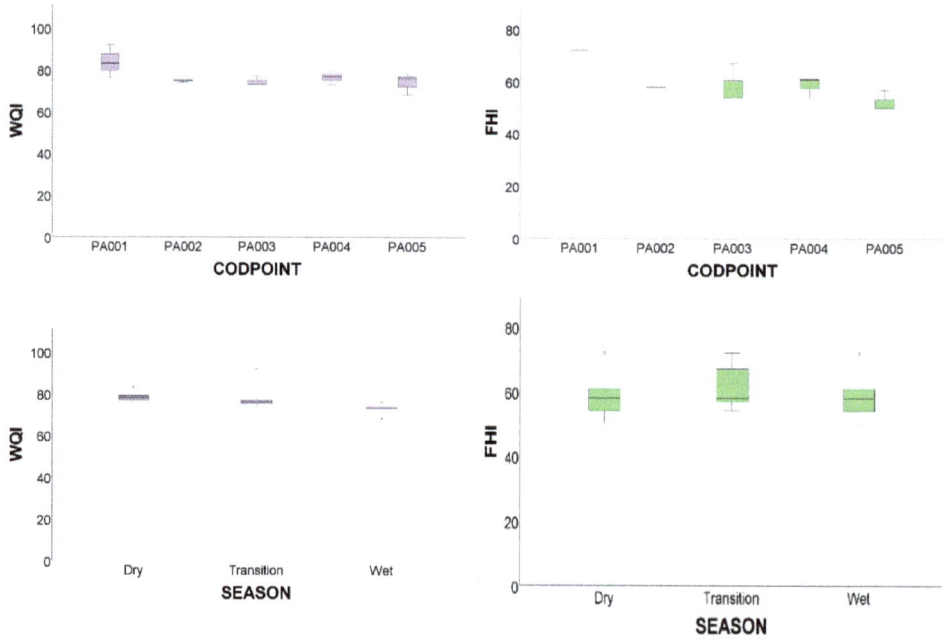

Figure 4: Boxplot of spatial-temporal variations of WQI and the FHI of Pacora River.

Table 4: Descriptive statistics and Kruskal–Wallis of WQI and the FHI of Pacora River.

Index	N	Min.	Max.	Mean	Std. error	Variance	Std. dev.	Median	SEASON p (same)	CODPOINT p (same)
WQI	15	68	92	76.6	1.4	29.4	5.422177	76	0.0375	0.3037
FHI	15	50	72	59.86667	1.966061	57.98095	7.614522	58	0.8842	0.04627

3.3.2 Fluvial habitat index

The FHI values are presented in Table 1, which shows that for the points PA001WE, PA001DR and PA001TR they present values from 72 to 86 corresponding to the very good ratings. For points PA 002WE, PA 002DR, PA 002TR, PA 003WE, PA003DR, PA 003TR, PA004WE, PA004DR, PA004TR, PA005WE, PA005DR, PA005TR they present values of 50 to 67 corresponding to the good ratings.

3.3.2.1 Spatio-temporal analysis of FHI

3.3.2.1.1 Spatial analysis

If there are significant differences between points PA001, PA002, PA003, PA004, PA005 according to the Kruskal–Wallis Test as shown in Fig. 4. The minimum FHI values were presented at point PA005 and the maximum values at point PA001 as described in Table 4.

3.3.2.1.2 Temporal analysis

The temporal variations are presented in Fig. 4, according to the Kruskal–Wallis Test there are no significant differences between the dry, wet and transitional seasons in the FHI.

According to ANAM [8], the Pacora River has an acceptable rating, this allows the development and consumption for agricultural, recreational and commercial activities, as well as the presence of aquatic life.

4 CONCLUSION

The results obtained in this study indicate that high areas of the Matasnillo, Juan Diaz and Pacora Rivers have the best ratings in the WQI and in FHI, decreasing their score in the middle and low areas.

In the application of the Kruskal–Wallis Test for the spatial analysis of the WQI in Matasnillo and Juan Diaz Rivers there are significant differences between the sampling sites, except in the Pacora River, for the case of the temporal analysis of the WQI for the Matasnillo and Juan Diaz Rivers there are no significant differences except in Pacora River.

The application of the Kruskal–Wallis Test for the spatial analysis of the FHI in the Matasnillo and Juan Diaz Rivers there are significant differences between the sampling sites except in Pacora River, for temporal analysis of the FHI in the Matasnillo, Juan Díaz and Pacora Rivers there are no significant differences between the wet, dry and transitional season.

Low WQI and FHI scores indicate the environmental impact caused by inadequate management of solid and liquid waste from human settlements, commercial and industrial activities that put pressure on these urban basins.

ACKNOWLEDGEMENTS

This research was financed by the Scholarship of the Subprogram of Doctoral and Postdoctoral Scholarships of the National Secretariat of Science and Technology (SENACYT) in conjunction with the Institute for the Training and Use of Human Resources (IFARHU). This research is part of the project Environmental Impact of Multiple Stressors in Aquatic Ecosystems of the Metropolitan Area of Panama, financed by SENACYT.

REFERENCES

[1]   Miserendino, M.L., Brand, C. & di Prinzio, C.Y., Assessing urban impacts on water quality, benthic communities and fish in streams of the Andes mountains, Patagonia (Argentina). *Water Air and Soil Poll.*, **194**, pp. 91–110, 2008.

[2]   Mamun, Md. & An, K.-G., The application of chemical and biological multi-metric models to a small urban stream for ecological health assessments. *Ecological Informatics*, **50**, pp. 1–12, 2019.

[3]   Aazami, J., Esmaili-Sari, A., Abdoli, A., Sohrabi, H. & Van den Brink, P., Monitoring and assessment of water health quality in the Tajan River, Iran using physicochemical, fish and macroinvertebrates indices. *Journal of Environmental Health Science and Engineering*, pp. 13–29, 2015.

[4]   Leprieur, F., Beauchard, O., Blanchet, S., Oberdorff, T. & Brosse, S., Fish invasions in the world's river systems: when natural processes are blurred by human activities. *PLoS Biology*, **6**(2), e28, 2008.

[5]   Magoulick, D.D., Dekar, M.P., Hodges, S.W., Scott, M.K., Rabalais, M.R. & Bare, C.M., *Scientific Reports Nature Portfolio*, **11**, 10704, 2021.

[6]   Poff, N.L., Richter, B.D., Arthington, A.H., Bunn, S.E., Naiman, R.J., Kendy, E.,
       Acreman, M., Aspe, C., Bledsoe B.P., Freeman, M.C., Henriksen, J., Jacobson, R.B.,
       Kennen, J.G., Merritt D.M., O'Keefee, J.H., Olden, J.D., Rogers, K., Tharme, R.E. &
       Warner, A., The ecological limits of hydrologic alteration (ELOHA): A new
       framework for developing regional environmental fow standards. *Freshwater
       Biology*, **55**, pp. 147–170, 2009.
[7]   Belmar, O., Velasco, J. & Martinez-Capel, F., Hydrological classification of natural
       fow regimes to support environmental fow assessments in intensively regulated
       Mediterranean Rivers, Segura River Basin (Spain). *Environmental Management*, **47**,
       pp. 992–1004, 2011.
[8]   ANAM (National Environmental Authority), Report on the monitoring of water
       quality in the watersheds of Panama, 2009–2012.
[9]   APHA (American Public Health Association), *Standard Methods for the Examination
       of Water and Wastewater*, 22nd ed., American Public Health Association: Washington,
       DC, 2012.
[10]   Pardo, I. et al., El hábitat de los ríos mediterráneos. Diseño de un índice de diversidad
       de hábitat. *Limnetica*, **21**(3–4), pp. 115–133, 2002.

ASSESSMENT OF SURFACE AND GROUNDWATER CONTAMINATION AND SEASONAL VARIATION AT THE TANNERY AREA IN DHAKA, BANGLADESH

ASIA AKTER[1], AFROSE SULTANA CHAMON[2],
MD. NADIRUZZAMAN MONDOL[2] & SYED MOHAMMED ABUL FAIZ[2]
[1]Department of Plant Sciences, University of Idaho, USA
[2]Department of Soil, Water and Environment, University of Dhaka, Bangladesh

ABSTRACT

Hazaribagh in Dhaka is home to about 150 footwear factories and is known as a contamination hub. Although some tanneries have relocated to new industrial parks, it has been seen that many are still running. In the present study, 42 water samples, including surface and groundwater, from six different sampling points of the Hazaribag tannery complex were examined for one year (December 2019–August 2020; dry and rainy seasons respectively). This research aimed to assess the surface and groundwater quality of the sampling area and the seasonal variation on different parameters. During sampling, a sequence was maintained, from the tannery source points to the downstream Buriganga River, the ultimate destination of tannery effluents. Chemical and physicochemical properties pH, DO, BOD, NO_3^-, C^{2+}, Mg^{2+}, K^+, Na^+, Cl^-, TDS, TSS, EC, Eh, and turbidity of wastewater were evaluated using the standard method. However, our primary attention was on heavy metal contamination. Trace metals were analyzed by atomic absorption spectrophotometer. The excessively higher concentration of heavy metals was observed in the first three sampling sites. It was also found that heavy metal concentrations decreased significantly from the source point to the downstream Buriganga River. Such variations in metal content with the location may have occurred due to increasing distance from the source point and percolation of wastewater to downward groundwater by soil. Comparatively lower concentrations of heavy metals were observed in the rainy season than dry season due to dilution with rainwater but still was above the maximum allowable concentration. Concentrations of all investigated heavy metals and some physicochemical parameters such as pH, DO, BOD, TDS, TSS, EC, Cl^-, and Na^+ of all the surface and groundwater samples were found above the national (DoE) and international (WHO) standards.

Keywords: tannery industries, heavy metal pollution, surface water, groundwater water pollution, tannery effluents.

1 INTRODUCTION

Leather processing is one of the most promising sectors of Bangladesh. Among 220 tanneries in Bangladesh, about 90–95% of all tanneries are located in Hazaribagh, Dhaka [1]. Twenty-five hectares of tannery complex are in Dhaka municipality's western part and inside the flood protection dam. About 60,000 tons of raw hides and skins are processed here per year [2] and generate 7.7 and 88 million tons of liquid and solid waste [3]. These wastes are not managed in efficient ways. As a result, this area has now become a chemically contaminated hub. An average of 30–35 m^3 of wastewater is produced per rawhide processing per ton [4]. Wastewater is discharged into nearby open drains, ponds, and lagoons. It makes its way to the other side of the dam and ultimately flows into the river Buriganga, which is considered Dhaka's lifeline. People in and around the city, directly and indirectly, depend on Buriganga for a different purposes. Tannery wastewater and effluents are the main culprits of extreme biodiversity loss in the river, and the color of river water turns black. After decades of discussion on the topic, the government launched a relocation project to move the tannery from Hazaribag to the north of Dhakar near Savar, where all the tanneries would share a central effluents treatment plant (CETP). Many tanneries have been relocated in the newly formed leather industry [5]. The main goal of relocating this industry was to make rivers safe from pollution. However, the goal is not attained in the newly formed industrial area [6]. As the Central effluent treatment plant has not been built, the same hazard will achieve in the area. They are now discharging non-treated effluent near the Dhaleshawri River [7]. Tanning is a process by which rawhide and skin turn into soft and durable leather. It is time-consuming, and many chemicals and mechanisms are involved here. About 40 heavy metals and acids are used during the soaking, tanning, and post tanning processing of hides and skins [8], including sodium sulfite, basic chromium sulfate, various painting agents, bactericides, soda ash, CaO, ammonium sulfide, ammonium chloride, sodium bisulfate, Cl, H_2SO_4, formic acid, sodium formate, sodium bicarbonate, vegetable tannins, system, resins, polyurethane, dyes, fat emulsions, pigments, binders, waxes, lacquers and formaldehyde and enzymes [9]. Only 20% of chemicals used in the tanning process are absorbed by leather rest are released as waste [10]. In the future area of Hazaribag will be used for commercial and residential purposes. However, soil and water, including groundwater, will suffer from tannery wastes [11]. A recent study observed a very high concentration of heavy metals in the surrounding area's soil and plant samples [12]. There is an urgent need to evaluate surface and groundwater in Hazaribag.

2 MATERIALS AND METHOD

2.1 Surface and groundwater sampling

Six sampling points were chosen based on the assumption of pollution intensity. The first three spots were located inside, and the rest three were outside the flood protection embankment of Dhaka. Spot I was considered the main disposal point of tanneries, and the increasing numbers of the spots indicant ate increasing distance from the main disposal point. Water samples were collected from the study area in two phases, December 2019 (dry season) to August 2020 (rainy season). The sampling spots were kept fixed throughout the whole sampling period. The sampling points were geo-referenced with GPS (global positioning system) and marked on the map. GPS locations of sampling points are given in Table 1. The location map of the study area and sampling location is shown in Figs 1 and 2. Water samples of 500 ml (with three replications) were collected from each site in sterilized plastic bottles fitted with a liquid-tight stopper. Groundwater samples were collected from deep tube well (DTW) (>150 m), located at sampling point IV. The water samples were immediately acidified with 4 ml of concentrated hydrochloric acid (HCl) per liter and analyzed within seven days of collection.

Table 1: GPS location of sampling point (both dry and rainy seasons).

I	23044.013'N	90021.807'E
II	23041.156'N	90021742'E
III	23044.552'N	90021.604'E
IV + GW	23044.673'N	90021.549'E
V	23044.600'N	90021.279'E
VI	23044.501'N	90021.109'E

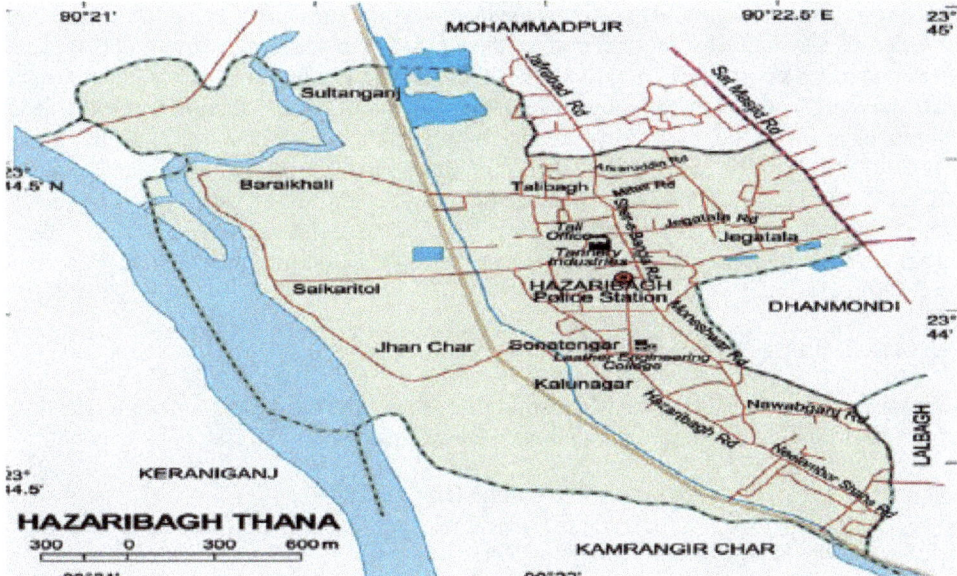

Figure 1: Location map of study area [16].

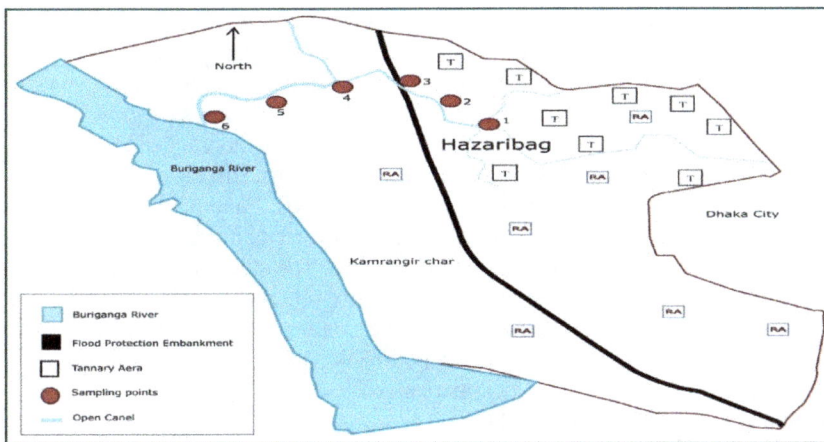

Figure 2: Sampling site at Hazaribag tannery area.

2.2 Analytical procedure

Determination of pH for all samples was done using the HI-2211 (Hanna Instrument). DO and BOD were measured with the electrolytic method by using an electric probe (Hanna HI 98193). Determination of chloride was done by the titration method. The micro Kjeldahl distillation method measured nitrates (NO_3^-) in water [13]. Potassium (K) was determined turbidimetrically after digesting the water sample with the $HNO_3 \cdot HClO_4$ (1:3) mixture by a spectrophotometer at 769 nm [14]. Sodium (Na) and potassium (K) were determined by atomic absorption spectrophotometer (AAS). Calcium (Ca) and magnesium (Mg) were measured titrimetrically by EDTA (ethylenediaminetetraacetic acid) method. Determination of chloride was done by the titration method [15]. Total dissolved solids (TDS) were measured using a portable TDS (HI98301 DiST1) meter. Total soluble salts (TSS) in water were measured gravimetrically by evaporating 20 ml of water sample at 1100°C and expressing it in mg kg^{-1}. Electrical conductivity (EC) was measured using an EC meter (HI98301 DiST1). The redox potential (Eh) of water samples was determined redox meter (PCE PH-30). The Turbidity of the water sample was determined by using a turbidity meter (LUTRON TU-2016). Heavy metals (Cr, Zn, Pb, Cd, Mn, Fe, and Ni) were analyzed by using an atomic absorption spectrophotometer (model no AA421) following the method prescribed by APHA [15].

2.3 Statistical analyses

Data analysis was done using the Microsoft Office Excel 2007 and SPSS software. The tests of significance of different sampling points were calculated by DMRT at a 5% level.

3 RESULT AND DISCUSSION

3.1 Chemical and physicochemical parameters

The chemical and physicochemical properties of wastewater are analyzed and presented in Table 2. The pH of the water sample was found to be in the range of 6.23 to 7.83 and 6.77 to 8.15 for dry and rainy seasons respectively and, was within the standard regulatory limits [17], [18]. There was no significant variation in pH values during the dry and rainy seasons. This also conforms to previous works [3]. A maximum of 8.15 pH was recorded during the rainy phase at sampling point III. The pH of water is greatly influenced by the concentration of CO_2 [19]. A very low value of DO was observed at various sampling points of the Hazaribag tannery area. The DO across multiple sampling points ranged from 0.00 to 0.45 and 0.65 to 0.96 mg/L in the dry season and rainy season respectively and in the dry season lowest DO was observed at the sampling point, I (0.00 mg/L), and 0.65 was observed at the same sampling point in the rainy season. The groundwater had a higher DO level in both seasons; 3.90 and 1.2 mg/L in two consecutive seasons. The study of the wastewater quality at the different sampling spots during the dry and rainy seasons revealed that tannery wastes heavily polluted the water and the Buriganga river water, so much, so that the DO of the wastewater samples was found to be nil during the dry season and no fish and other aquatic life were found living, up to 500 m downstream of the sluice gate. DO in all samples was detected below the standard limit [17], [18].

BOD values were higher in the rainy season 217 mg/L in relation to the dry season 150 mg/L. This might be due to warmer temperatures in that season enhanced microbial activities, consuming more oxygen to decompose industrial and tannery effluents and domestic wastes, which are drained out into the river by monsoon rain and surface run-off. The inversely proportional relationship between BOD and DO was also observed throughout this study. However, the important fact is that the BOD values were high, which reduced the amount of oxygen (DO) available to fish and rendered the water body unsuitable for the fishery. Similar works were also reported earlier [20]. BOD in all sampling points exceeds the recommended maximum allowable concentration [17], [18].

NO_3^- in the sample, a ranged from 96 to 189 and 12 to 185 mg/L for the dry and rainy seasons, respectively. Compare to the dry season a lower value of NO_3^- was observed in the

Air and Water Pollution XXX 103

Table 2: Wastewater's chemical and physicochemical properties in dry and rainy phases.

Spot No	pH	DO	BOD	NO$_3^-$	Ca	Mg	K	Na	Cl	TDS	TSS mg/50 mL	EC dS/cm	Eh mV	Turbidity FTU
					mg/L									
Dry season														
I	6.23	0.00	150	156	4.3	1.2	0.3	98	789	3,160	0.16	2.1	378	4,096
II	7.83	0.45	138	163	4.1	2.1	0.6	100.2	895	2,860	0.09	7.4	365	674
III	7.73	0.35	128	125	4.5	2.6	0.2	95	789	2,580	0.09	6.10	363	563
IV	7.61	0.33	124	147	4.9	1.9	0.2	79	656	2,320	0.03	3.87	353	800
V	7.35	0.25	112	96	3.6	1.3	0.6	16.2	548	643	0.04	2.64	344	347
VI	7.02	0.38	168	189	3.8	1.9	0.1	23.2	458	592	0.02	6.74	197	60
GW	7.36	3.9	0	121	0.2	0.03	0.01	9	18	125	0.01	0.228	110	3.68
Rainy season														
I	8.07	0.65	212	125	4.5	2.3	0.2	126	568	486	1.2	7.23	325	530
II	7.59	0.96	206	185	4.6	2.1	0.2	158	698	430	0.23	8.01	125	442
III	8.15	0.89	194	163	5.6	2.03	0.2	63	456	329	0.56	7	389	166
IV	7.11	0.69	154	73	6.2	2.56	0.2	48	678	226	0.48	3.72	321	41.5
V	7.24	0.85	114	45	4.2	1.3	0.3	69	458	116	0.12	2.85	378	27.4
VI	7.16	0.74	217	184	3.8	1.8	0.6	128	325	223	0.45	6.42	296	13.6
GW	6.77	1.2	0	12	0.34	0.05	0.01	2	11	105	0.00	0.84	241	2.4
FAO [17]	6–9		30	50						2,100	30			550
DoE [18]	6–9	4.5	50	50						2,100	50	1.2		550

WIT Transactions on Ecology and the Environment, Vol 259, © 2022 WIT Press
www.witpress.com, ISSN 1743-3541 (on-line)

rainy season. The highest value in the dry season might be due to the continuous accumulation of tannery effluents. All NO_3^- values of DTW during the rainy season were obtained were within regulatory limits of 50 mg/L [18].

Ca^{2+} in wastewater at different sampling points of the Hazaribag tannery is ranged from 0.2 to 4.9 and 0.34 to 6.2 mg/L for the dry and rainy seasons, respectively. Mg^{2+} ranged from 0.03 to 2.6 and 0.05 to 2.56 mg/L for the dry and rainy seasons. Na^+ in wastewater at various sampling points ranged from 9 to 100.2 and 2 to 158 mg/L for the dry and rainy seasons. Compared to the dry season a higher value of Ca^{2+}, Mg^{2+} and Na^+ were observed in the rainy season. However, potassium, calcium, and magnesium values recorded in this study were found within the recommended regulatory limit in water <10 mg/L [17], [18]. Very high Sodium concentration was recorded in different samples of this study except for the groundwater. Similar findings were also reported in a previous article [9]. Chloride in the investigated wastewater varied from 18 to 895 and 11 to 698 mg/L in dry and rainy seasons respectively. However, chloride, except for DTW values (18 and 11 mg/L) recorded in this study, was found to be above the recommended regulatory limit in water >250 mg/L [21]. Tannery effluents and municipal sewage waste might be responsible for this contamination. The sharp fall of Chloride concentration in the rainy season in the wastewater was due to dilution of rainwater by monsoon rain or less discharge of tannery effluents from industries at the time of sampling. The effect of chloride in an aquatic environment does not seem to have received much attention. However, a certain high concentration of any ion is bound to affect overall biota drastically. TDS in wastewater at different samples ranged from 125 to 3,160 and 2.4 to 530 mg/L for dry and rainy seasons. However, the TDS of the first three sampling points in the dry phase exceeds the regulatory value of 2,100 mg/L [18]. The variations in TDS concentrations in wastewater are due to the discharge of effluents and wastes from the tannery industries. Comparative higher values were in the dry season than the rainy season. Lower TDS values obtained during the rainy season could result from several physicochemical reactions such as sedimentation, coagulation, fixation, oxidation, and precipitation [22]. The number of solids present in a water body can be used to estimate the pollution load due to both organic and inorganic contaminants present.

TSS ranged from 0.0 to 0.16 and 0.00 to 1.2 mg/50 mL, respectively, in the dry and rainy seasons. In the dry season, maximum TSS was observed at sampling point I (0.16 mg/50 ml). Gradually decreasing values were observed from sources point to downstream, outside the barrier. The lowest turbidity value was observed in tube well water (0.00 mg/50 ml). Similar findings were observed in the rainy season. TSS value of the wastewater samples at different spots decreased in the dry season. The subsequent increase was observed in the rainy season, indicating a large amount of waste discharged from tannery industries and heavy monsoon rainfall causing surface runoff from the surrounding area during the investigation period. EC indicates the total concentration of the ionized constituents of water and is usually correlated closely with TDS [19]. EC in wastewater at various sampling points ranged from 0.228 to 7.4 and 0.84 to 8.01 dS/m, respectively, in the dry and rainy seasons. EC of all samples was found within the regulatory limit of 1,200 dS/cm [17].

The redox potential characterizes the oxidation-reduction state of water. Eh, the value of the water sample in two seasons was determined on the 28th day after sampling and ranged from −436 to −110 and −389 to −241 mV in dry and rainy seasons, respectively. The lowest redox potential was recorded in the rainy season at −389 mV. The highest one was recorded during the dry season at −110 mV from samplings point I and groundwater, respectively. A low redox potential in the rainy season might be due to changes in the composition of organic matter and metal complexes in the overlying water and higher redox values in the dry season, probably due to the continuous turnover and movement of the waters. Values of EC, Na, Ca,

K, Mg and Eh indicate high wastewater pollution even in the Buriganga River (sampling point VI), where water is not stagnant. Many authors also reported similar findings [23], [24].

The turbidity in various sampling points ranged from 3.68 to 4096 and 2.4 to 530 FTU, respectively, in the dry and rainy seasons. Turbidity is the opposite phenomenon of transparency and has an inverse relationship. Turbidity is essential because of the aesthetic consideration and the organism's pathogenicity that can be hidden on tiny colloidal particles [25]. In the dry season, maximum values were observed at sampling point I, 4096 FTU, and gradually decreasing values from the source point to downstream, outside the barrier. The lowest value of turbidity was observed in groundwater (3.68 FTU). Similar findings were observed in the rainy season, where the highest turbidity value was honored at the source point, i.e., the sampling point I, 530 FTU. Turbidity above the regulatory limit was observed at sampling spots II, III, IV, and I in the dry season [17].

3.2 Heavy metals

Cr in wastewater samples ranged from 4,609.8 to 8,995.0 mg/L and 17.86 to 863.03 mg/L in dry and rainy seasons, respectively (Table 3). In the dry season, highest 8,995.0 mg/L was observed at sampling point I, i.e., the main disposal point of tanneries. Various chromium salts used in tannery industries, or vast amounts of wastewater may contaminate spot I and effluents are now continuously added from other sectors [23]. Significant differences were found between sampling points II, III, IV, and V, VI. DMRT calculated the significance tests of different samples at the 5% level. In the rainy season, a comparatively lower concentration of Cr was observed at the same points. 17.48 and 863.03 mg/L was the lowest and highest observation of Cr at sampling points V and I. This downfall may be due to the dilution of water by rainfall. Imamul-Huq [9] reported 0.28 ppm Cr in wastewater in the Hazaribag area. Similar findings were also reported by Nuruzzaman et al. [26]. Cr in all samples, both dry and rainy seasons across the MAC (maximum allowable concentration) for inland water d 0.5 mg/kg [18] and drinking water of WHO 0.05 mg/kg [21].

Table 3:   Chromium (mg/L) in wastewater at sampling points of study area both in dry and the rainy season.

Spot No	\multicolumn{4}{c}{Chromium (mg/L)}			
	Dry season	Rainy season	National Standards DoE [18]	WHO Standards WHO [17]
I	8,995.0 a	863.03 a		
II	5,967.50 b	674.56 ab		
III	5,619.17 b	62.90 b		
IV	6,232.66 b	275.18 ab		
V	3,644.08 d	39.51 b	0.5	0.05
VI	4,609.8 c	17.86 b		
GW	28.33	2.14 b		

Zinc in water samples in the dry season was 3.10 mg/L at sampling points I, 2.6, 2.21, 1.76, 1.74, 1.73, and 0.601 mg/L in sampling points II, III, IV, V, VI, and groundwater, respectively (Table 4). There were significant differences among different sampling spots except for points II and III. In the case of the rainy season, 2.46 and 0.25 ppm was the highest and lowest observation of Zn that was found at sampling point VI, and I. Ullah et al. [23] reported 0.12 ppm of Zn in wastewater at the Hazaribag area. Similar findings were also reported [26]. All samples crossed the MAC for inland and drinking water 5 mg/kg [18], [21].

Table 4: Zinc (mg/L) in wastewater at sampling points of the study area both in dry and rainy seasons.

Spot No	Dry season	Rainy season	National Standards DoE [18]	WHO Standards WHO [17]
I	17.38 a	16.91 a		
II	17.33 b	12.82 b		
III	17.67 b	13.76 b		
IV	22.16 c	16.91 b		
V	26.03 c	14.49 bc	5	5
VI	31.01c	24.63 bc		
GW	6.1 d	2.51 c		

In the dry season, 1.53 and 3.15 mg/L was the lowest and highest observation of Pb at sampling points I and III (Table 5). No Significant differences were found among different sampling spots. 0.366 to 0.94 mg/L was the lowest and highest findings of Pb that were observed at sampling points I and II. Pb in all samples was cross the traditional value of effluents for inland surface water, 0.1 mg/kg [18].

Table 5: Lead (mg/L) in wastewater at sampling points of study area both in dry and rainy season.

Spot No	Dry season	Rainy season	National Standards DoE [18]	WHO Standards WHO [17]
I	3.15 a	0.70 b		
II	3.04 a	0.94 a		
III	2.39 a	0.44 c		
IV	2.23 a	0.51 c		
V	1.53 a	0.36 c	0.1	0.01
VI	1.59 a	0.51 c		
GW	1.63 a	0.78 ab		

Cd in wastewater samples ranged from 0.007 to 0.035 and 0.005 to 0.035 mg/L, respectively, in dry and rainy seasons (Table 6). Groundwater and VI samples were significantly different from other spots. A comparative lower concentration of Cd was observed at the same sampling point in the rainy season. Maybe rainfall was responsible for this mitigation. Cd in all observed samples crosses the national standards for inland 0.05 mg/kg [18], and the maximum allowable drinking water concentration by WHO is 0.005 mg/kg [21].

Mn 1.8 mg/L was the highest observation of Mn found at sampling point I, i.e., the main disposal point of tanneries and located inside the embankment (Table 7). Spot I may contaminate by various manganese salts used in tannery industries. Statically this sample and groundwater were significantly different from other samples. In the rainy season, 0.03 and 1.73 mg/L was the lowest and highest concentration of Mn observed at sampling point I and groundwater. Nuruzzaman et al. [26] reported 0.014 ppm Mn in wastewater at the Hazaribag tannery area. Similar findings were also reported by Ullah et al. [23]. Mn concentration at six sampling points during both dry and rainy seasons across the national and international standards for inland water 5 mg/kg [18] and drinking water (0.05 mg/kg) [21].

Table 6: Cadmium (mg/L) in wastewater at sampling points of study area both in dry and rainy season.

Spot No	Dry season	Rainy season	National Standards DoE [18]	WHO Standards WHO [17]
			Cadmium (mg/L)	
I	0.035 a	0.035 a		
II	0.033 a	0.020 bc		
III	0.028 a	0.025 b		
IV	0.030 a	0.005 d		
V	0.026 a	0.015 c	0.05	0.005
VI	0.007 b	0.016 bc		
GW	0.006 b	0.005 d		

Table 7: Manganese (mg/L) in wastewater at sampling points of study area both in dry and rainy season.

Spot No	Dry season	Rainy season	National Standards DoE [18]	WHO Standards WHO [17]
			Manganese (mg/L)	
I	12.81 c	7.3 a		
II	18.31 b	3.33 a		
III	21.38 a	0.20 c		
IV	12.98 c	0.59 b		
V	11.81 c	0.05 d	5	5
VI	7.15 d	0.45 b		
GW	4.48 e	0.03 d		

Fe in wastewater samples ranged from 38.93 to 275.01 and 32.11 to 113.27 mg/L in dry and rainy seasons, respectively (Table 8). Spot I was significantly different from other spots except and III. A comparative lower value of Fe was observed at the same sampling point in the rainy season. The previous investigation reported 0.14 ppm Fe in wastewater in the same area [23]. Similar findings were also reported by Nuruzzaman et al. [26]. Fe in all samples was above the national and international standers.

Table 8: Iron Chromium (mg/L) in wastewater at sampling point of study area both in dry and rainy season.

Spot No	Dry season	Rainy season	National Standards DoE [18]	WHO Standards WHO [17]
			Iron mg/L	
I	275.01 a	113.27 a		
II	226.43 a	81.6 b		
III	104.76 a	68.27 b		
IV	61.68 b	61.6 b		
V	42.60 b	54.94 c	2	0.3
VI	38.93 b	32.11 bc		
GW	13.60 c	4.61 d		

Ni in wastewater during the dry season was 1.93 at sampling points I, 0.84, 0.51, 0.34, 1.64, 0.84, and 0.51 mg/L in sampling points II, III, IV, V, VI, and groundwater, respectively (Table 9). No significant differences were found among different samples. But the situation was different in the case of the rainy season. Significant differences were found among other sampling spots. 0.341 and 1.57 ppm was the lowest and highest concentration of Ni observed at sampling point V and I. Ni in all samples were above the MAC for drinking water (0.1 mg/kg) [21] and inland water bodies 1 mg/kg [18].

Table 9: Nickel chromium (mg/L) in wastewater at sampling point of study area both in dry and rainy season.

| Spot No | Nickel (mg/L) | | | |
	Dry season	Rainy season	National Standards DoE [18]	WHO Standards WHO [17]
I	1.93 a	1.57 a		
II	1.64 a	0.80 ab		
III	0.84 a	0.502 ab		
IV	0.51 a	0.341 b		
V	0.34 a	0.65 ab	1	0.1
VI	0.84 a	0.67 ab		
GW	0.51 a	0.11 b		

4 CONCLUSION

In the study, in the area of Hazaribagh, the industrial wastewater is discharged into the river Buriganga via open canals and low lands with extremely high concentrations of Na^+, K, Mg^{2+}, Ca^{2+}, NO_3^- Cl^- and heavy metals, especially Cr, Zn, Pb, Cd, Mn, Fe, and Ni. The high organic matter content was also detected, as evident in high BOD and low DO observe this study's high BOD and low DOD, NO_3^-, Ca^{2+}, Mg^{2+}, K^+, Na^+, Cl^-, TDS, TSS, EC, Eh, and turbidity negatively affect aquatic life and the ecosystem. There was an unusually high concentration of toxic metals in the first three sampling sites. Heavy metal concentrations also decreased dramatically from the source point to the Buriganga River downstream. Such variations in metal content with the location may have occurred due to increasing distance from the source point and percolation of wastewater to downward groundwater by soil. Due to dilution with rainwater, heavy metal concentrations were lower in the rainy season than in the dry season, but they were still above the maximum permitted concentration. The attention of all investigated heavy metals and some physicochemical parameters such as pH, DO, BOD, TDS, TSS, EC, Cl, and Na^+ of all the surface and groundwater samples were found above the DoE and WHO standards. There is the possibility of further contamination of surface and groundwater if untreated tannery wastes and effluents are not considered. It is an emergence to relocate all tanneries to the new industrial sites to reduce pollution. Our last sampling point was the Buriganga River, where concentrations of heavy metals in river water have also exceeded the maximum allowable concentration of DoE and WHO for drinking water due to tanning activities. Surface Water of the study area of Hazaribagh tannery industries is said to be highly polluted, especially with cation, anion, organic matter, and especially with heavy metals, so it is not suitable for the external purposes of human activities such as bathing, washing, irrigation, and ground water is highly polluted, especially with heavy metals, so it is not recommended as a drinking water for human consumption.

ACKNOWLEDGEMENTS

The work was supported by the National Science and Technology (NST) Fellowship (grant number 164); Ministry of Science and Technology, Bangladesh. Experiments were conducted in different laboratories of the Soil, Water and Environment Department, University of Dhaka, and Bangladesh Council of Scientific and Industrial Research (BCSIR).

We are grateful to the head of the SWED department and the chairperson of BCSIR for their continuous inspiration. We want to express our cordial thanks to tannery workers, producers, and traders in the Hazaribag area for their cooperation, suggestion, and assistance with sample collection.

REFERENCES

[1]   Paul, H.L., Antunes, A.P.M., Covington, A.D., Evans, P. & Philips, P.S., Bangladeshi leather industry: An overview of recent sustainable developments. *Journal of the Society of Leather Technologists and Chemists*, 2013.

[2]   Rasul, M.G., Faisal, I. & Khan, M.M.K., Environmental pollution generated from process industries in Bangladesh. *Int. J. Environ. Pollut.*, 2006. DOI: 10.1504/IJEP.2006.010881.

[3]   Bhuiyan, M.A.H., Suruvi, N.I., Dampare, S.B., Islam, M.A., Quraishi, S.B., Ganyaglo, S. & Suzuki, S., Investigation of the possible sources of heavy metal contamination in lagoon and canal water in the tannery industrial area in Dhaka, Bangladesh. *Environ. Monit. Assess.*, 2011. DOI: 10.1007/s10661-010-1557-6.

[4]   Islam, B.I., Musa, A.E., Ibrahim, E.H., Sharafa, S.A.A. & Elfaki, B.M., Evaluation and characterization of tannery wastewater. *J. For. Prod. Ind.*, 2014.

[5]   Bari, M.L., Simol, H.A., Khandoker, N., Begum, R. & Sultana, U.N., Potential human health risks of tannery waste-contaminated poultry feed. *J. Heal. Pollut.*, 2015. DOI: 10.5696/2156-9614-5-9.68.

[6]   Sarker, M.N.I., Wu, M., Liu, R. & Ma, C., Challenges and opportunities for information resource management for e-governance in Bangladesh, 2019. DOI: 10.1007/978-3-319-93351-1_53.

[7]   Khan, F.E., Jolly, Y.N., Islam, G.M.R., Akhter, S. & Kabir, J., Contamination status and health risk assessment of trace elements in foodstuffs collected from the Buriganga River embankments, Dhaka, Bangladesh. *Int. J. Dhaka C. Bangladesh*, 2014. DOI: 10.1186/s40550-014-0001-z.

[8]   Buljan, J., Cost of tanned waste treatment. United Nations Industrial Development Organization, 2005.

[9]   Imamul-Huq, S.M., Critical environmental issues relating to tanning industries in Bangaladesh. *ACIAR Proc.*, 1998.

[10] McCartor, A., Becker, D., Hanrahan, D., Ericson, B., Thomen, A., Fuller, R., Jones, D. & Caravanos, J., World's worst pollution problems report 2010. *Environment*, 2010.

[11] Halim, M.A., Sumayed, S.M., Majumder, R.K., Ahmed, N., Rabbani, K. & Karmaker, S., Study on groundwater, river water and tannery effluent quality in southwestern Dhaka, Bangladesh: Insights from multivariate statistical analysis. *J. Nature, Sci. Sustain. Technol.*, 2011.

[12] Mondol, M., Asia, A., Chamon, A. & Faiz, S., Contamination of soil and plant by the Hazaribagh tannery industries. *J. Asiat. Soc. Bangladesh, Sci.*, **43**, pp. 207–222, 2017. DOI: 10.3329/jasbs.v43i2.46518.

[13] Egli, H., Kjeldahl guide, Büchi Labortechnik AG: CH-9230 Flawil, Switzerland, 2008.

[14] Jackson, A.P. & Alloway, B.J., The transfer of cadmium from agricultural soils to the human food chain. *Biogeochem. Trace Met.*, **1**, pp. 109–158, 1992.

[15] APHA, AWWA & WEF, Standard methods for examination of water and wastewater. 2012. DOI: 10.5209/rev_ANHM.2012.v5.n2.40440.

[16] Siddiqee, M.H., Islam, M.S. & Rahman, M.M., Assessment of pollution caused by tannery-waste and its impact on aquatic bacterial community in Hajaribag, Dhaka. *Stamford J. Microbiol.*, **2**, pp. 20–23, 2013. DOI: 10.3329/sjm.v2i1.15208.

[17] FAO, General standards for discharge of effluents, 2014.

[18] DoE, Guide for assessment of effluent treatment plants, pp. 9–10, 2008.

[19] Boyd, C.E. & Lichtkoppler, F., Water quality management in pond fish culture. *Research and Development Series., Int. Cent. Aquac.*, **22**, pp. 1–30, 1979.

[20] Khan, M.A.I., Hossain, A.M., Huda, M., Islam, M.S. & Elahi, S.F., Physico-chemical and biological aspects of monsoon waters of ashulia for economic and aesthetic applications: Preliminary studies. *Bangladesh J. Sci. Ind. Res.*, 1970. DOI: 10.3329/bjsir.v42i4.747.

[21] WHO, *Guidelines for Drinking-water Quality*, 3rd ed., Vol. 1. Recommendations, 2004. DOI: 10.1016/S1462-0758(00)00006-6.

[22] Wasserman, G.A., Liu, X., Parvez, F., Ahsan, H., Levy, D., Factor-Litvak, P., Kline, J., van Geen, A., Slavkovich, V., Lolacono, N.J., Cheng, Z., Zheng, Y. & Graziano, J.H., Water manganese exposure and children's intellectual function in Araihazar, Bangladesh. *Environ. Health Perspect.*, 2006. DOI: 10.1289/ehp.8030.

[23] Ullah, S.M., Nuruzzaman, M. & Gerzabek, M.H., Heavy metal and microbiological pollution of water and sediments by industrial waters, effluents and slums around Dhaka City. *Tropical Limnology*, Vol. 3, Tropical rivers, wetlands and special topics, 1995.

[24] Chowdhury, F.J., S.M.I.H. & M.A.I., Accumulation of various pollutants by some aquatic macrophytes found in the Buriganga River. *Proceeding of the 25th Bangladesh Science Conference*. DOE 1992. Training Manual on Environmental Management, pp. 121–145, 1996.

[25] Vesilind, P.A., Peirce, J.J. & Weiner, R.F., Measurement of water quality. *Environmental Pollution and Control*, 1990. DOI: 10.1016/b978-0-409-90272-3.50008-9.

[26] Nuruzzaman, M., Gerzabek, M.H., Islam. A., Rashid, M.H. & Ullah, S.M., *Bangladesh Journal of Soil Science*, **25**, pp. 1–10, 1998.

WATER MANAGEMENT IN COLOMBIA FROM THE SOCIO-ECOLOGICAL SYSTEMS FRAMEWORK

MIGUEL A. DE LUQUE-VILLA & MAURICIO GONZÁLEZ-MÉNDEZ
Departamento de Ecología y Territorio, Facultad de Estudios Ambientales y Rurales,
Pontificia Universidad Javeriana, Colombia

ABSTRACT
Nowadays the main challenge for water management is to seek water resilience. The socio-ecological systems approach was developed by Ostrom with the aim of synthesizing knowledge to foster a better understanding of the relationship between people, institutions and the environment. The SES framework allows us to identify the SES main characteristics and provide information on the modes of interaction and self-organization processes between the actors involved in the collective management of the common use resource. Actually, Colombia presents a great vulnerability of water scarcity, so the purpose of this paper was to review actual water management in Colombia and propose the social-ecological systems framework that allows safeguarding and sustaining the water cycle, guaranteeing a sufficient water supply and providing a stable climate system for secure human well-being in Colombia.
Keywords: Colombia, socio-ecological systems, water management, water resilience.

1 INTRODUCTION

Water availability per capita worldwide is steadily decreasing due to the growing world population in relation to available water [1]. The main causes of the deficit in water supply are climate change and climate variability, which determine variations in the amount and temporal distribution of precipitation affecting the hydrological cycle, contributing to water scarcity [2]–[4]. Recent studies have also argued that water demand contributes to water scarcity, with agriculture being the main consumer [5]–[8]. Proof of this is the drastic transformation of land cover, which has led to a conflict between the temporal and spatial distribution of water resources, as well as an intensification of the contradiction between supply and demand, which seriously restricts the sustainable development of a basin [9].

Domínguez et al. [10] studied the Colombian water supply and potential water demand by the different productive sectors, concluding that the maximum demand and supply magnitudes do not coincide in space, causing conflict and high levels of pressure on the water resource. According to the National Water Study 2018 [11], the region with the highest vulnerability to water deficit in Colombia is the Caribbean hydrographic area, the second hydrographic area with the highest demand in the country. Likewise, the study forecasts shortages in the departments of La Guajira, Magdalena, Cesar and Bolivar, where at least 50% of their municipalities will be affected. According to the report of the United Nations Development Programme [12].

Sustainable water management is a growing concern worldwide, Baudoin and Arenas [13] conducted a bibliometric study of current water resource management worldwide and found that it is being approached from different theoretical approaches and that there is a risk that water management research is going in different directions simultaneously, without academics taking advantage of each other's work. Evidencing that management theories do not align toward an improved understanding of sustainable water management. The main objective in this paper was to review water management in Colombia and propose a social-ecological systems framework that allows the water resilience.

2 SOCIAL-ECOLOGICAL SYSTEMS (SES) FRAMEWORK

The socio-ecological systems (SES) approach was developed by Ostrom [14], [15] and refined by McGinnis and Ostrom [16] with the aim of synthesizing knowledge to foster a better understanding of the relationship between people, institutions and the environment [17]. The SES framework was developed as a tool to study the relationships between the multiple levels that compose a common, its application can support the identification of the main characteristics of the SES and provide information on the modes of interaction and the processes of self-organization between the actors involved in the collective management of the common pool resources [18]. Fig. 1 and Table 1 show the first-tier components and second-tier variables of the SES framework

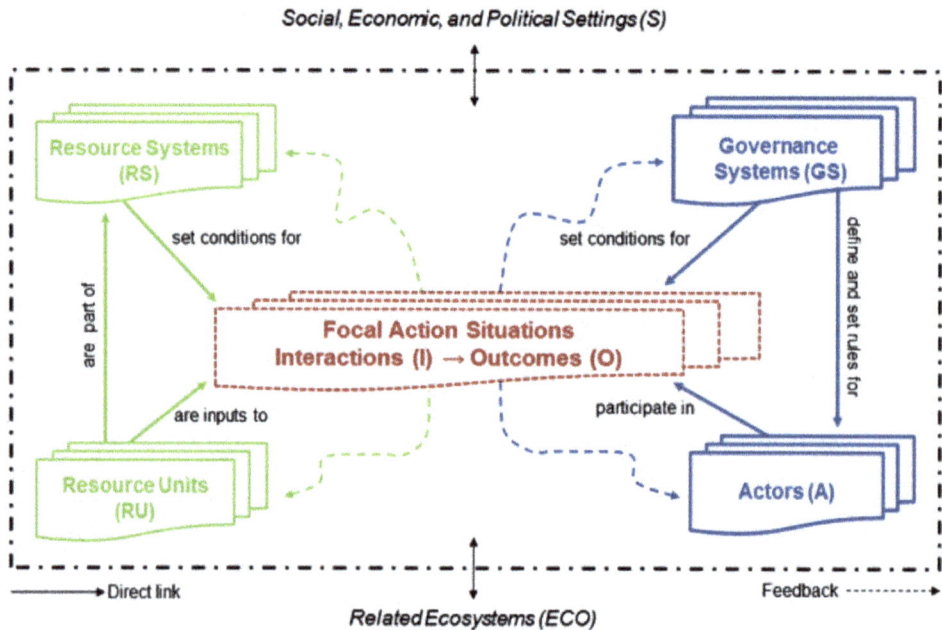

Figure 1:   Revised social-ecological system (SES) framework with multiple first-tier components [16].

Application of the SES framework to particular cases requires a three-step process. In the first step, the analyst must select a focal level of analysis by answering such questions as: What types of interactions and outcomes related to a particular resource system (or group of systems) and related resource units (or other relevant goods and services) are most relevant to my analytical or diagnostic concerns? What types of actors are involved? Which governance systems influence the behavior of these actors? [18].

3 COLOMBIA WATER MANAGEMENT

The objective of the water resource management Colombian policy is to guarantee the sustainability of water resources, through efficient and effective management and use, articulated with land use, planning and the conservation of ecosystems that regulate the water supply, considering water as a factor of economic development and social welfare, and

Table 1: Second-tier variables of a social-ecological system [16].

First-tier variable	Second-tier variables
Social, economic, and political settings (S)	S1 – Economic development
	S2 – Demographic trends
	S3 – Political stability
	S4 – Other governance systems
	S5 – Markets
	S6 – Media organizations
	S7 – Technology
Resource systems (RS)	RS1 – Sector (e.g., water, forests, pasture, fish)
	RS2 – Clarity of system boundaries
	RS3 – Size of resource system
	RS4 – Human-constructed facilities
	RS5 – Productivity of system
	RS6 – Equilibrium properties
	RS7 – Predictability of system dynamics
	RS8 – Storage characteristics
	RS9 – Location
Governance systems (GS)	GS1 – Government organizations
	GS2 – Nongovernment organizations
	GS3 – Network structure
	GS4 – Property-rights systems
	GS5 – Operational-choice rules
	GS6 – Collective-choice rules
	GS7 – Constitutional-choice rules
	GS8 – Monitoring and sanctioning rules
Resource units (RU)	RU1 – Resource unit mobility
	RU2 – Growth or replacement rate
	RU3 – Interaction among resource units
	RU4 – Economic value
	RU5 – Number of units
	RU6 – Distinctive characteristics
	RU7 – Spatial and temporal distribution
Actors (A)	A1 – Number of relevant actors
	A2 – Socioeconomic attributes
	A3 – History or past experiences
	A4 – Location
	A5 – Leadership/entrepreneurship
	A6 – Norms (trust-reciprocity)/social capital
	A7 – Knowledge of SES/mental models
	A8 – Importance of resource (dependence)
	A9 – Technologies available

Table 1: Continued.

First-tier variable	Second-tier variables
Action situations: Interactions (I) → Outcomes (O)	I1 – Harvesting
	I2 – Information sharing
	I3 – Deliberation processes
	I4 – Conflicts
	I5 – Investment activities
	I6 – Lobbying activities
	I7 – Self-organizing activities
	I8 – Networking activities
	I9 – Monitoring activities
	I10 – Evaluative activities
Related ecosystems (ECO)	O1 – Social performance measures (e.g., efficiency, equity, accountability, sustainability)
	O2 – Ecological performance measures (e.g., overharvested, resilience, biodiversity, sustainability)
	O3 – Externalities to other SESs Related ecosystems (ECO)
	ECO1 – Climate patterns
	ECO2 – Pollution patterns
	ECO3 – Flows into and out of focal SES

implementing processes of equitable and inclusive participation [19]. The above was divided into several objectives focused on supply, demand, quality, risk, institutional strengthening and governance. However, in 2019 the Office of the Comptroller General of the Republic [20] concluded that three years after finalizing the implementation of the policy, there are weaknesses and problems, the main ones being difficulties in the study of supply and demand. The water resource information system to evaluate supply is not clear, the environmental authorities are unaware of the real demand for water resources in their territories, and finally, the actions aimed at protecting water resources are not coordinated with planning and land use instruments. The Institute of Hydrology, Meteorology and Environmental Studies (IDEAM) is in charge of carrying out the integral evaluation of water resources in Colombia through the National Water Study, which is presented every 4 years. The conceptual framework of the National Water Study is the hydrological cycle, which is based on the water balance to estimate the natural amount of renewable freshwater available and its spatial and temporal distribution according to the behavior of precipitation, evapotranspiration and runoff variables in the hydrographic subzones that make up the Colombian territory [11]. From the socio-ecological systems approach, it is evident that the study is focused on the system and the resource units, not taking into account the variables of the social, economic and political environment, the governance systems, action situations and related ecosystems.

The following is a bibliographic review of water resource management in Colombia, for which a search was conducted in both Spanish and English. The search covered the years 2000–2022 in the google academic, ScienceDirect and Scopus databases. Journal articles and doctoral theses were reviewed. To perform the analysis, each article was reviewed from two approaches. The first consisted of evaluating the factual characteristics, theoretical framework and methodology, in order to identify the line of research. The second consisted of a qualitative evaluation of the content using the social-ecological systems approach, in

Table 2: Descriptive overview of the research stream and analysis through the social-ecological systems framework.

Articles	Article type	Research stream	Analysis through the social-ecological systems framework
Climate change and water resources in Colombia [21].	Research	Spatial water availability modeling	Resource systems (RS), Resource units (RU), Related ecosystems (ECO)
Reflection on the vision of integrated management of vulnerability due to water shortages in rural zone of Colombia [22]	Review	Analysis of the concepts of adaptive management, resilience and governance for integrated water resource management	Governance systems (GS), Related ecosystems (ECO)
Integration of hydrological and economical aspects for water management in tropical regions. Case study: Middle Magdalena Valley, Colombia [23]	Research	Spatial and temporal water availability modeling and Integrated water resource management	Social, economic, and political settings (S), Concern, resource systems (RS), Resource units (RU), Action situations: Interactions (I), Related ecosystems (ECO)
Integrated water resource management in Colombia: Paralysis by analysis? [24]	Review	Integrated water resource management	Governance systems (GS)
Water management analysis in the Magdalena basin in Colombia [25]	Research	Spatial and temporal water availability and water demand modeling	Social, economic, and political settings (S), Concern, resource systems (RS), Resource units (RU), Action situations: Interactions (I), Related ecosystems (ECO)
Collective action for watershed management: Field experiments in Colombia and Kenya [26]	Research	Collective action on basins	Resource systems (RS), Governance systems (GS), Actors (A), Action situations: Interactions (I) → Outcomes (O)
Development and testing of a river basin management simulation game for integrated management of the Magdalena-Cauca river basin [27]	Research	Serious game for basin integrated management	Resource systems (RS), Governance systems (GS), Resource units (RU), Actors (A), Action situations: Interactions (I) → Outcomes (O)
Relaciones demanda-oferta de agua y el índice de escasez de agua como herramientas de evaluación del recurso hídrico Colombiano [10]	Research	Spatial and temporal water availability and water demand modeling	Resource systems (RS), Resource units (RU), Related ecosystems (ECO)

Table 2: Continued.

Articles	Article type	Research stream	Analysis through the social-ecological systems framework
Validation of the three-step strategic approach for improving urban water management and water resource quality improvement [28]	Research	Water resource quality improvement	Action situations: Interactions (I) → Outcomes (O)
Considerations of environmental ethics in the comprehensive management of water resources in the Quindío River basin [29]	Review	Environmental ethics in the integrated management of water resources	Resource systems (RS), Resource units (RU), Related ecosystems (ECO)
Components to formulate an integrated water resources management. Case study in Quindío Basin [30]	Research	Elements for formulating integrated water resource management	Resource systems (RS), Resource units (RU), Action situations: Interactions (I) → Outcomes (O), Related ecosystems (ECO)
Climate variability, climate change and water resources in Colombia [31]	Review	Climate variability and climate change	Resource systems (RS), Resource units (RU), Related ecosystems (ECO)
Integrated water resource management as a strategy for adaptation to climate change [32]	Review	Integrated water resource management	Resource systems (RS), Resource units (RU), Related ecosystems (ECO)
Challenges in environmental governance: An approach to the implications of integrated water resource management in Colombia [33]	Review	Water governance	Governance systems (GS)
A case study of group decision method for environmental foresight and water resources planning using a fuzzy approach [34]	Research	Fuzzy opinion to solve the decision-making problem on water management	Resource systems (RS), Resource units (RU), Action situations: Interactions (I), Related ecosystems (ECO)
Research advances on the integral management of water resource in Colombia [35]	Research	Integrated water resource management	Resource systems (RS), Resource units (RU)
Water resource management and economic value [36]	Review	Management and economic value of water resources	Resource units (RU)

Table 2: Continued.

Articles	Article type	Research stream	Analysis through the social-ecological systems framework
Planning and management of water resources: A review of the importance of climate variability [37]	Review	Climate variability	Resource systems (RS)
Simulation of infrastructure options for urban water management in two urban catchments in Bogotá, Colombia [38]	Research	Stormwater harvesting, reuse of industrial waters, water-saving technology in residential sectors, and reuse of water from washing machines	Resource systems (RS), Resource units (RU)
Engaged universals and community economies: The (human) right to water in Colombia [39]	Research	Right to water	Governance systems (GS)
Development and implementation of a water-safety plan for drinking-water supply system of Cali, Colombia [40]	Research	Water safety plan	Resource systems (RS), Resource units (RU), Governance systems (GS)
El índice de escasez de agua ¿Un indicador de crisis ó una alerta para orientar la gestión del recurso hídrico? [41]	Research	Indicators of water supply and demand	Resource systems (RS), Resource units (RU)
Comparative analysis of integrated water resources management models and instruments in South America: Case studies in Brazil and Colombia [42]	Review	Integrated water resource management	Governance systems (GS)
Framework for water management in the food–energy–water (FEW) nexus in mixed land-use watersheds in Colombia [43]	Research	Food–energy–water nexus framework	Resource systems (RS), Resource units (RU)
Governance of water resources in Colombia: between progress and challenges [44]	Review	Water resources governance	Governance systems (GS)

order to develop an integral perspective of water resource management in the research. Table 2 shows the descriptive overview of the research stream and analysis through the social-ecological systems framework.

We collected 25 articles in total. The articles collected are predominantly research articles with 60% (15). The remaining 40% (10) were literature reviews. The main focus of the research streams were the spatial and temporal water availability and water demand modeling and the Integrated water resource management.

The conclusion of all the articles that studied the Integrated water resource management is that in Colombia it has had a limited application and no spaces have been developed to promote the development of collective water management projects.

From the analysis through the social-ecological systems framework it is evident that in Colombia most of the studies are focused on the Resource systems (RS), Resource units (RU) and Governance systems (GS). Therefore, although these studies on water management cover a part of the problem, there are other unknown variables that have the power to explain better how the socio-ecological system works, so it is necessary to rethink a conceptual framework from all the first-tier components, that will allow to achieve water resilience.

4 COLOMBIA WATER MANAGENMENT SOCIO-ECOLOGICAL SYSTEM FRAMEWORK

The idea that anthropogenic water systems can be resilient to the pressures of scarcity only through economic decision-making (e.g. cost–benefit analysis), engineering and technology is beginning to be undermined [45]. This framework has not previously been applied within the realm of water resources management and may prove a useful tool in planning for better management and, ultimately, future resilience [45]. Governance of water functions for social-ecological resilience must be continuous, adaptive, iterative and incremental in nature to enable learning in a dynamically changing environment, to be nested and harmonized across sectors and scales to manage interactions and interconnections [46].

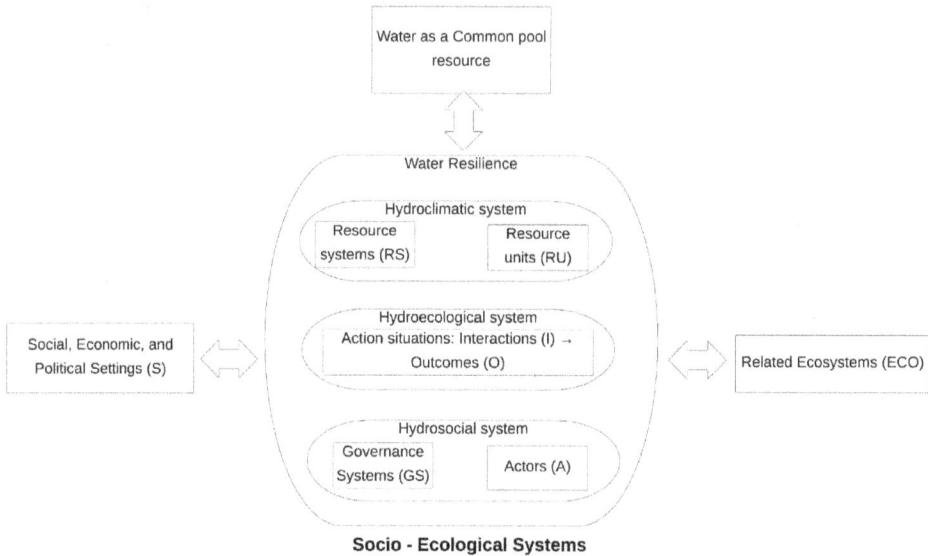

Figure 2:   Colombian social-ecological system (SES) framework. *(Source: Adapted from [16].)*

The central topic addressed in the Colombia water management SES framework is the water as a common pool resource. The water as a common pool resource can be viewed as a SES because it entails a human group that is dependent on a resource, an ecosystem that provides it and a socio-cultural system that allows humans interact with the ecosystem.

In Fig. 2 the water as a common pool resource is linked to a socio-ecological systems box where are the main subsystems of a SES and illustrates how the different subsystems occur. These interactions generate outcomes that feedback the SES' subsystems. The SES box includes the concept of water resilience because this is a notion that includes the dynamic perspective to the SESs analysis.

5 CONCLUSIONS

As a result of the review of the state of the art of current water management in Colombia, it is observed that the integrated water resource management approach has not been implemented in the best way. This is due to the fact that all the studies are only focused on some variables that affect the system and the problem of water resilience is not being seen from the socio-ecological system framework.

A new conceptual framework for water management in Colombia was proposed, involving a larger number of variables to better understand the interaction between hydroclimatic, hydrosocial and hydroecological systems. This new l framework is important because water management had never been approached in Latin America from the socio-ecological systems framework.

ACKNOWLEDGEMENT

The authors thank the Pontificia Universidad Javeriana for all of their support for the project.

REFERENCES

[1] Lakshmi, V., Fayne, J. & Bolten, J., A comparative study of available water in the major river basins of the world. *J. Hydrol.*, vol. 567, no. March, pp. 510–532, 2018, DOI: 10.1016/j.jhydrol.2018.10.038.

[2] DeNicola, E., Aburizaiza, O.S., Siddique, A., Khwaja, H. & Carpenter, D.O., Climate change and water scarcity: The case of Saudi Arabia. *Ann. Glob. Heal.*, **81**(3), pp. 342–353, 2015. DOI: 10.1016/j.aogh.2015.08.005.

[3] Gampe, D., Nikulin, G. & Ludwig, R., Using an ensemble of regional climate models to assess climate change impacts on water scarcity in European river basins. *Sci. Total Environ.*, **573**, pp. 1503–1518, 2016. DOI: 10.1016/j.scitotenv.2016.08.053.

[4] Dinar, S., Katz, D., De Stefano, L. & Blankespoor, B., Do treaties matter? Climate change, water variability, and cooperation along transboundary river basins. *Polit. Geogr.*, **69**, pp. 162–172, 2019. DOI: 10.1016/j.polgeo.2018.08.007.

[5] Winter, J.M., Lopez, J.R., Ruane, A.C., Young, C.A., Scanlon, B.R. & Rosenzweig, C., Representing water scarcity in future agricultural assessments. *Anthropocene*, **18**, pp. 15–26, 2017. DOI: 10.1016/j.ancene.2017.05.002.

[6] Carvalho, A.P.P., Lorandi, R., Collares, E.G., Di Lollo, J.A. & Moschini, L.E., Potential water demand from the agricultural sector in hydrographic sub-basins in the southeast of the state of São Paulo-Brazil. *Agric. Ecosyst. Environ.*, **319**, 2021. DOI: 10.1016/j.agee.2021.107508.

[7] Schwaller, C., Keller, Y., Helmreich, B. & Drewes, J.E., Estimating the agricultural irrigation demand for planning of non-potable water reuse projects. *Agricultural Water Management*, **244**, 2021. DOI: 10.1016/j.agwat.2020.106529.

[8]   Bwambale, E., Abagale, F.K. & Anornu, G.K., Smart irrigation monitoring and control strategies for improving water use efficiency in precision agriculture: A review. *Agricultural Water Management*, **260**, 2022. DOI: 10.1016/j.agwat.2021.107324.

[9]   Chen, D., Li, J., Yang, X., Zhou, Z., Pan, Y. & Li, M., Quantifying water provision service supply, demand and spatial flow for land use optimization: A case study in the YanHe watershed. *Ecosyst. Serv.*, **43**, 101117, 2020. DOI: 10.1016/j.ecoser.2020.101117.

[10] Domínguez, E.A., Rivera, H.G., Sarmiento, R.V. & Moreno, P., Relaciones demanda-oferta de agua y el índice de escasez de agua como herramientas de evaluación del recurso hídrico Colombiano. December, 2008.

[11] IDEAM, *Estudio Nacional del Agua 2018*, Bogotá, 2019.

[12] PNUD, El caribe colombiano frente a los objetivos de desarrollo del milenio (ODM). 2009. http://www.pnud.org.co/img_upload/33323133323161646164616461646164/LINEADEBASEODMCARIBE.pdf

[13] Baudoin, L. & Arenas, D., From raindrops to a common stream: Using the social-ecological systems framework for research on sustainable water management. *Organ. Environ.*, **33**(1), pp. 126–148, 2020. DOI: 10.1177/1086026618794376.

[14] Ostrom, E., A diagnostic approach for going beyond panaceas. *Proc. Natl. Acad. Sci. U. S. A.*, **104**(39), pp. 15181–15187, 2007. DOI: 10.1073/pnas.0702288104.

[15] Ostrom, E., A general framework for analyzing sustainability of social-ecological systems. *Science (80-)*, **362**, pp. 419–422, 2009. DOI: 10.5055/jem.2013.0130.

[16] McGinnis, M.D. & Ostrom, E., Social-ecological system framework: Initial changes and continuing challenges. *Ecol. Soc.*, **19**(2), 2014. DOI: 10.5751/ES-06387-190230.

[17] Epstein, G. et al., Advances in understanding the evolution of institutions in complex social-ecological systems. *Curr. Opin. Environ. Sustain.*, **44**, pp. 58–66, 2020. DOI: 10.1016/j.cosust.2020.06.002.

[18] Perrotti, D., Hyde, K. & Otero Peña, D., Can water systems foster commoning practices? Analysing leverages for self-organization in urban water commons as social–ecological systems. *Sustain. Sci.*, **15**(3), pp. 781–795, 2020. DOI: 10.1007/s11625-020-00782-1.

[19] Ministerio de Ambiente Vivienda y Desarrollo Territorial, *Política Nacional de Gestión integal del Recurso Hídrico*, Bogotá, D.C., Colombia, 2010.

[20] Contraloría general de la República, Evaluación de la implementación de la política nacional para la gestión del recurso hídrico con énfasis en la oferta y la demanda 2015–2018, Colombia, 2019.

[21] Alarcón Hincapié, J., Zafra Mejía, C. & Echeverri Prieto, L., Climate change and water resources in Colombia. *Rev. U.D.C.A. Actual. Divulg. Científica*, **22**(2), pp. 1–10, 2019.

[22] Andrade, A., Osorio-Garcés, C.E. & Martinez-Idrobo, J.P., Reflection on the vision of integrated management of vulnerability due to water shortages in rural zone of Colombia. *Ing. Y Compet.*, **21**(2), pp. 1–14, 2019. DOI: 10.25100/iyc.v21i2.8342.

[23] Arenas Bautista, M.C., Integration of hydrological and economical aspects for water management in tropical regions. Case study: Middle Magdalena Valley, Colombia. 2020. https://repositorio.unal.edu.co/handle/unal/77944.

[24] Blanco, J., Integrated water resource management in Colombia: Paralysis by analysis? *Int. J. Water Resour. Dev.*, **24**(1), pp. 91–101, 2008. DOI: 10.1080/07900620701747686.

[25] Bolivar Lobato, M.I., Water management analysis in the Magdalena basin in Colombia. 2021. https://ediss.sub.uni-hamburg.de/handle/ediss/9179.

[26]   Cardenas, J.C., Rodriguez, L.A. & Johnson, N., Collective action for watershed management: Field experiments in Colombia and Kenya. *Environ. Dev. Econ.*, **16**(3), pp. 275–303, 2011. DOI: 10.1017/S1355770X10000392.

[27]   Craven, J., Angarita, H., Corzo Perez, G.A. & Vasquez, D., Development and testing of a river basin management simulation game for integrated management of the Magdalena-Cauca river basin. *Environ. Model. Softw.*, **90**, pp. 78–88, 2017. DOI: 10.1016/j.envsoft.2017.01.002.

[28]   Galvis, A., Van der Steen, P. & Gijzen, H., Validation of the three-step strategic approach for improving urban water management and water resource quality improvement. *Water (Switzerland)*, **10**(2), pp. 1–18, 2018. DOI: 10.3390/w10020188.

[29]   García Reinoso, P.L. & Obregón, N., Considerations of environmental ethics in the comprehensive management of water resources in the Quindío River basin. *Entramado*, **8**(2), pp. 12–37, 2012.

[30]   García Reinoso, P.L. & Obregón, N., Components to formulate an integrated water resources management. Case study in Quindio Basin. *Rev. Tecnol.*, **10**(2), pp. 73–83, 2011.

[31]   García, M.C., Piñeros Botero, A., Bernal Quiroga, F.A. & Ardila Robles, E., Climate variability, climate change and water resources in Colombia. *Rev. Ing.*, **36**, pp. 60–64, 2012. DOI: 10.16924/revinge.36.11.

[32]   García-González, M.L., Carvajal-Escobar, Y. & Jiménez-Escobar, H., Integrated water resource management as a strategy for adaptation to climate change. *Ing. Y Compet.*, **9**(1), pp. 19–29, 2007. DOI: 10.1080/03081067508717092.

[33]   González, N., Challenges in environmental governance: An approach to the implications of integrated water resource management in Colombia. *CienciaPolítica.*, **12**(23), pp. 205–229, 2017.

[34]   Halabi, A.X., Montoya-Torres, J.R. & Obregón, N., A case study of group decision method for environmental foresight and water resources planning using a fuzzy approach. *Gr. Decis. Negot.*, **21**(2), pp. 205–232, 2012. DOI: 10.1007/s10726-011-9269-z.

[35]   Hernández Pasichana, S.M. & Posada Arrubla, A., Research advances on the integral management of water resource in Colombia. *Rev. U.D.C.A Actual. Divulg. Científica*, **21**(2), pp. 553–563, 2018. DOI: 10.31910/rudca.v21.n2.2018.1079.

[36]   Munevar, W.G.D., Water resource management and economic value. *Rev. Finanz. y Polit. Econ.*, **7**(2), pp. 279–298, 2015. DOI: 10.14718/revfinanzpolitecon.2015.7.2.4.

[37]   Ortiz, A., Ruiz, M. & Rodríguez, J., Planning and management of water resources: A review of the importance of climate variability. *Rev. Logos, Cienc. Tecnol.*, **9**(1), pp. 100–105, 2017. https://www.redalyc.org/pdf/5177/517754057010.pdf.

[38]   Peña-Guzmán, C.A., Melgarejo, J., Lopez-Ortiz, I. & Mesa, D.J., Simulation of infrastructure options for urban water management in two urban catchments in Bogotá, Colombia. *Water (Switzerland)*, **9**(11), 2017. DOI: 10.3390/w9110858.

[39]   Perera, V., Engaged universals and community economies: The (human) right to water in Colombia. *Antipode*, **47**(1), pp. 197–215, 2015. DOI: 10.1111/anti.12097.

[40]   Pérez-Vidal, A., Escobar-Rivera, J.C. & Torres-Lozada, P., Development and implementation of a water-safety plan for drinking-water supply system of Cali, Colombia. *Int. J. Hyg. Environ. Health*, **224**, 113422, 2020. DOI: 10.1016/j.ijheh.2019.113422.

[41]   Posada, C.C., Domínguez, E. & Rivera, H.G., El índice de escasez de agua ¿Un indicador de crisis ó una alerta para orientar la gestión del recurso hídrico? *Rev. Ing. Fac. Ing. Univ. los Andes*, pp. 104–111, 2005.

[42] Rojas Padilla, J.H., Perez Rincón, M.A., Malheiros, T.F., Madera Parra, C.A., Prota, M.G. & Dos Santos, R., Comparative analysis of integrated water resources management models and instruments in South America: Case studies in Brazil and Colombia. *Rev. Ambient. e Agua – An Interdiscip. J. Appl. Sci.*, **8**(1), pp. 73–97, 2013. DOI: 10.4136/1980-993X.

[43] Torres, C., Gitau, M., Lara-Borrero, J. & Paredes-Cuervo, D., Framework for water management in the food-energy-water (FEW) nexus in mixed land-use watersheds in Colombia. *Sustain.*, **12**(24), pp. 1–27, 2020. DOI: 10.3390/su122410332.

[44] Zamudio Rodríguez, C., Governance of water resources in Colombia: Between progress and challenges. *Rev. Gestión y Ambient.*, **15**(3), pp. 99–112, 2012.

[45] Gittins, J.R., Hemingway, J.R. & Dajka, J.C., How a water-resources crisis highlights social-ecological disconnects. *Water Res.*, **194**, 116937, 2021. DOI: 10.1016/j.watres.2021.116937.

[46] Falkenmark, M. & Wang-Erlandsson, L., A water-function-based framework for understanding and governing water resilience in the Anthropocene. *One Earth*, **4**(2), pp. 213–225, 2021. DOI: 10.1016/j.oneear.2021.01.009.

SECTION 3
WATER TREATMENT

DESIGN OF THE INTERCEPTOR-COLLECTOR AND WASTEWATER TREATMENT SYSTEM FOR POLLUTION MITIGATION: A CASE STUDY

BETHY MERCHÁN-SANMARTIN[1,2,3], PAÚL CARRIÓN-MERO[1,2], FERNANDO MORANTE-CARBALLO[2,3,4], JOSUÉ BRIONES-BITAR[1,2], ADRIÁN GONZALEZ-RUGEL[5] & HAIRO VERA-DEMERA[5]
[1]Facultad de Ingeniería en Ciencias de la Tierra (FICT), ESPOL Polytechnic University, Ecuador
[2]Centro de Investigaciones y Proyectos Aplicados a las Ciencias de la Tierra (CIPAT),
ESPOL Polytechnic University, Ecuador
[3]Geo-Recursos y Aplicaciones (GIGA), ESPOL Polytechnic University, Ecuador
[4]Facultad de Ciencias Naturales y Matemáticas (FCNM), ESPOL Polytechnic University, Ecuador
[5]Independent consultant, Ecuador

ABSTRACT

Sustainable Development Goal 6 proposes clean water and sanitation for all. Urban parishes Veintimilla and Polibio Chávez of the Guaranda canton in Ecuador have a combined sewage network. However, they do not have an operational wastewater treatment plant (WTP), which discharges directly into the river generating significant pollution and affecting the health and well-being of the inhabitants. This work aims to design the interceptor-collector and treatment system for wastewater discharges to the Guaranda river, considering the technical aspects and sustainability criteria, contributing to the quality of life of its inhabitants, the mitigation of pollution, and the compliance with current regulations. The methodology used was: (i) review and analysis of existing information, which included topographical, social, economic, technical, and environmental restrictions; (ii) proposal and selection of alternatives; (iii) design of the selected alternative, with its respective budget, plans, and environmental analysis. The results were: (i) design errors in the existing system, industrial discharges contaminate domestic wastewater; (ii) alternatives were proposed that were assessed using the Likert scale, with number A1b being the winner; (iii) the estimated future population is 32,922 inhabitants, and the design period is 15 years, which generates a flow of 0.226 m³/s. The interceptor-collector consists of four bypass chambers; the WTP consists of two independent lines, each with preliminary treatment, a sand trap of $17.25 \times 3.85 \times 1.15$ m, and percolating filters $15.4 \times 7 \times 2.4$ m, and a clarifier of 15.4 m by 3.5 m in height. The budget is USD 941,953.49, including the environmental management plan.
Keywords: sustainability, pollution, sewage, percolating filters, wastewater treatment plant (WTP).

1 INTRODUCTION

Wastewater is defined as water that has been polluted due to human activities. These waters can be raw (grey or black) and can potentially cause significant damage to human and environmental health [1]. However, its use as a reliable and cost-effective pre-treated water source, particularly for agricultural or industrial applications, is increasingly recognised [2].

Water is the source most affected by population growth, so the need for effluent treatment has become a significant concern [3], [4]. However, the high costs of wastewater purification make it challenging to implement treatment plants, mainly due to the long distances that the effluent must be transported (often requiring pumping stations to reach the treatment area) [5]. Furthermore, the lack of low-cost and straightforward management technologies makes it impossible to use them in rural areas and communities with low population density. Due to this, research has been developed to optimise the effluent purification processes [6]–[8]. In a context like a case addressed in this study, an example was carried out in the Montañita community, Ecuador. Specifically, the application of a pilot plan for green filters to reduce the pollutants concentrations in wastewater [9]. This natural purification technique saves operating costs and makes it possible to contribute to reforestation with endemic trees, promoting the circular economy.

In general, many urban areas lack preventive systems against pollution by domestic wastewater [10], [11]. Globally, inadequate sanitation-wastewater treatment systems and a lack of reusable water affect small towns and rural communities [12]. As a result, 82% of the rural population in developing countries lack basic sanitation [13].

Ecuador is not a country alien to this problem. The National Water Secretariat (SENAGUA) revealed that approximately 20% of domestic wastewater is treated in the country. The rest is discharged directly into rivers and streams. As a result, 48.5% of the rural population has safe drinking water, and 86.3% has basic sanitation [14].

Guaranda city is part of the Ecuadorian highlands; being the capital of Bolivar province, it is positioned at an average elevation of 2,668 m.a.s.l. (meters above sea level). It is composed of three urban parishes. This city has a sewage system that began to be built in the 1960s and 1970s, which discharged directly into the Guaranda River. In 2001, a treatment plant was inaugurated to which the discharges from the Ignacio Veintimilla and Polibio Chávez parishes were conducted through an outfall collector. This was a participatory action between the municipality and small communities, which is very common in Ecuador [15]. Due to problems in its design, the collector collapsed due to the intense rains in this area, returning to the previous system (direct discharge to the river) [16].

This research aims to design the interceptor and treatment system for wastewater discharges from the Ignacio Veintimilla and Polibio Chávez parishes in Guaranda. In addition, consider the treatment system's technical aspects and sustainability criteria to guarantee water pollution mitigation and compliance with current regulations.

2 STUDY AREA

Guaranda city is the capital of the Bolivar province, located 220 km from Quito (Ecuador's capital) and 150 km from Guayaquil (the country's principal port). According to the National Statistics and Census Institute (INEC) [17], its population is 55,374 inhabitants. Guaranda is made up of three urban parishes: Guanujo, Ignacio Veintimilla and Polibio Chávez (see Fig. 1(a)). According to the National Meteorology and Hydrology Institute (INAMHI) [18], it has an average temperature of 13°C.

90% of Guaranda's urban population receives drinking water from the Municipal Potable Water and Sewage Company (EMAPA-G). However, the urban periphery and the rural area do not have drinking water services. Also, according to INEC [17], 31% of the population has a sewerage network connection, and another 31% uses a septic tank. Currently, no area of the city of Guaranda has wastewater treatment plants in operation.

There are some contribution zones corresponding to the Veintimilla and Polibio Chávez parishes. Wastewater is collected through household connections and rainwater through sinks in each area. The primary collectors lead these to the discharge points on the Guaranda River (see Fig. 1(b)).

3 MATERIALS AND METHODS

For the development of this research, three phases are proposed that include: (i) review and analysis of existing information, which includes topographical, social, economic, technical, and environmental restrictions; (ii) proposal and selection of alternatives; (iii) design of the selected alternative, with its respective budget, and environmental analysis (see Fig. 2).

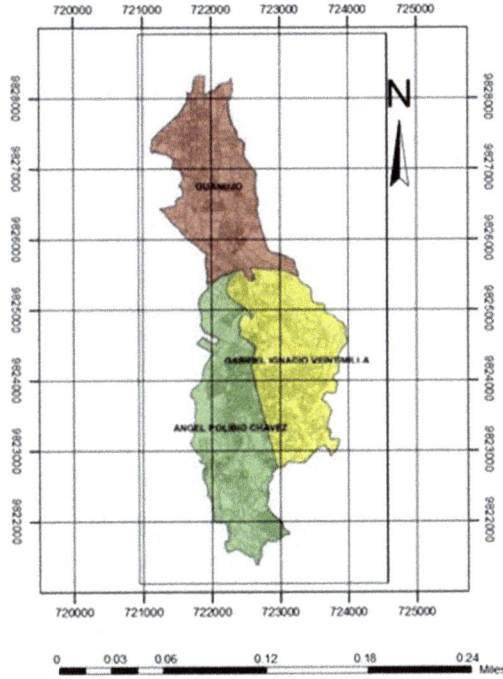

(a)

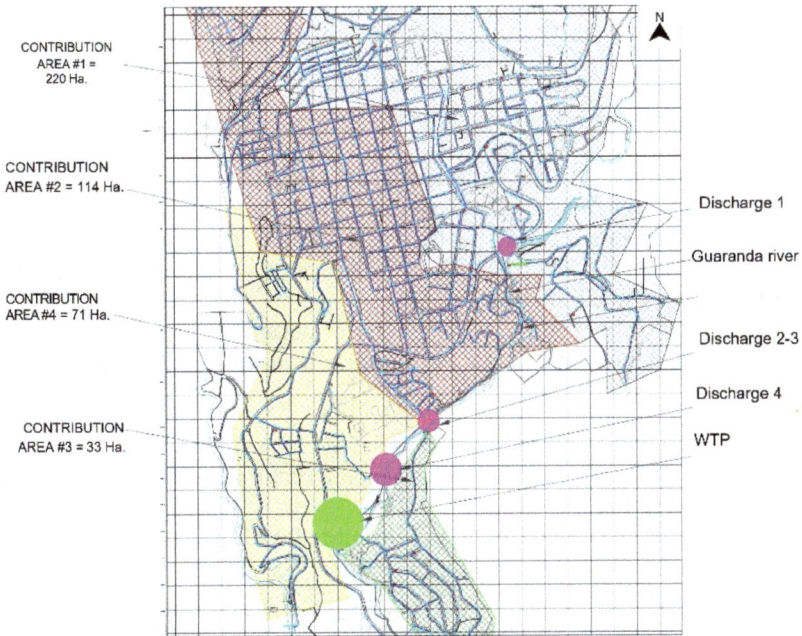

(b)

Figure 1:   (a) The geographical location of Guaranda city and its urban parishes; and
             (b) Wastewater and rainwater discharge points [16], [19].

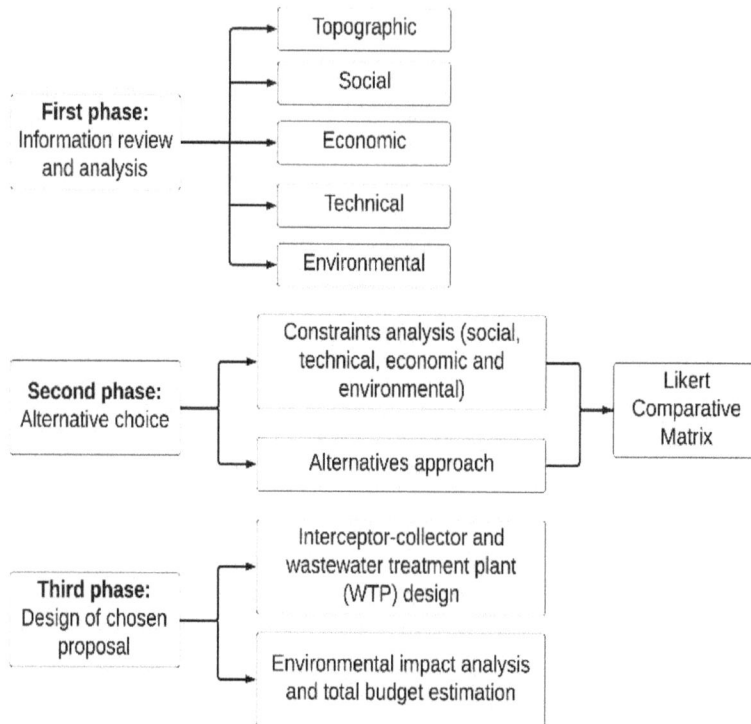

Figure 2: The methodological research map.

3.1 First phase: information review and analysis

The investigation begins with the information collected from the study area, including topographical, social, economic, technical, and environmental data. Data processing involves analysing and comparing existing information. In addition, a technical visit was made to the study area with EMAPA-G personnel, where they also provided files about the sewerage and drinking water master plan of the Guaranda canton.

The data collected in this phase are the basis for designing the wastewater collection, transport, and treatment system according to the study area with optimal operability.

3.2 Second phase: Alternative choice

In this phase, the analysis of the study area's economic, environmental, social, and technical restrictions was carried out.

In the economic analysis, from meetings with EMAPA-G, a system with low operation and maintenance costs (OPEX) is desired. In addition, a system that works by gravity and with this avoids energy expenses for pumping.

The environmental analysis considered whether the project would be in an area considered to be protected by forests or fauna, decreed by the Ministry of the Environment, Water and Ecological Transition (MAATE). Furthermore, it is required to make the most negligible environmental impact as far as possible. Therefore, laboratory tests were carried out on the discharge waters of Guaranda city (see Fig. 1). These were compared to national regulations such as the Unified Text of Secondary Environmental Legislation (TULSMA, its acronym in Spanish) and the Ecuadorian Practice Code [20].

In the social analysis, the surrounding community was considered since there is a possibility of generating bad odours and pests (e.g., mosquitoes in winter). Furthermore, depending on the final effluent quality, its use in agricultural areas in surrounding areas was also analysed.

The technical analysis analysed the territorial availability of the treatment system and the topography. Population growth was calculated for the design period. The population growth projection was carried out using three different methods (arithmetic, geometric and exponential) following the regulations with starting data taken from the censuses carried out by INEC [17].

Once the restrictions were analysed, alternatives were proposed for the interceptor and the purifying system for the study area. The Likert assessment method was used, for which it is taken from 1 for "very unfavourable" to 5 for "very favourable". The one with the highest score was selected as the most convenient or optimal.

3.3 Third phase: Design of the chosen proposal (interceptor-collector + purifying system)

The design of the interceptor-collector system was based on three main criteria: (i) consideration of the natural slope of the terrain prioritising transport by gravity, (ii) design of chambers with the necessary dimensions to contain wastewater and rainwater so that, autonomously, excess rainwater is diverted to the Guaranda river, and (iii) the water from the interceptor is directed to the wastewater treatment plant (WTP). In addition, the design considers the current regulations of the country and the conditions of the study area. The design period was projected at 15 years for the selected alternative to effectively comply with the transport of wastewater and minimise its implementation costs.

WTP design must comply with the maximum permissible limits (MPL) for treated effluents, stipulated in the current Ecuadorian regulations (TULSMA-2015 Book VI Annex 1 Table 10). Therefore, a pre-treatment system was designed that removes coarse material and prevents the accumulation of sand and grease that could compromise subsequent treatment. Additionally, installing a purification system based on a percolator filter and a clarifier is proposed. The main objective is the disinfection of wastewater, notably eliminating the organic load (see Fig. 3).

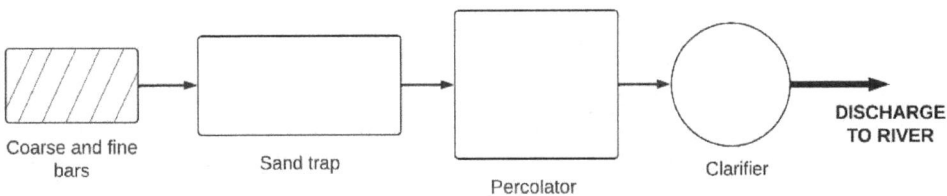

Figure 3: Scheme of the treatment system proposed.

In addition, the environmental impact assessment was carried out for the proposed design in the study area. This evaluation was carried out using the integrated relevant criteria (IRC) methodology proposed by Buroz [21]. It consists of assigning a numerical value to each impact based on indicators that make up the environmental impact value index (EIVI).

4 RESULTS

4.1 Population growth projection

In the last census carried out by the INEC in 2010, the population registered in the urban parishes of Polibio Chávez, and Gabriel Ignacio Veintimilla was 23,874 inhabitants. According to the CPE-INEN standard, the design period for a population such as Guaranda must be 15 years. Table 1 shows the projections for 2021, 2026, 2031 and 2036. In Fig. 4, it can see the results until the year 2036 (with the R^2 factor), so the projection was chosen by the arithmetic method because it has a factor $R^2 = 1$.

Table 1: Population projections of the Ignacio Veintimilla and Polibio Chávez parishes.

Year	Arithmetical method	Geometric method	Exponential method
2021	27,702	28,351	29,875
2026	29,442	30,655	33,081
2031	31,182	33,146	36,630
2036	32,922	35,840	40,561

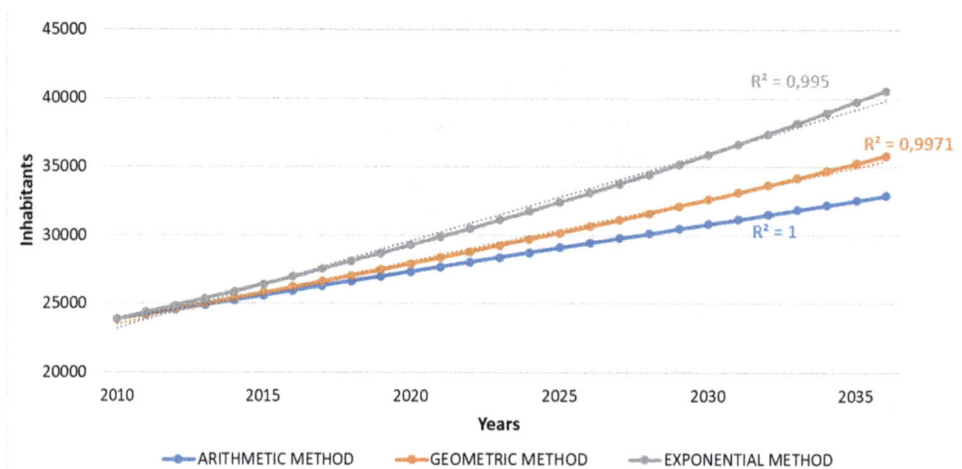

Figure 4: Population projection with its respective R^2 factor.

Also, for the population growth projection choice, the page of the Decentralized Autonomous Government of Guaranda (GAD-Guaranda, acronym in Spanish) was reviewed [22]. It indicates that the population, in 2021, was 25,000. Therefore, the population projection by the arithmetic method would be the closest, according to data from the GAD-Guaranda.

4.2 Laboratory results

The laboratory tests of residual water were carried out at four discharge points. These results were compared with the MPL of the TULSMA. Table 2 shows the main results, underlining the parameters that exceed the TULSMA limits in red.

Table 2: Laboratory tests carried out on each wastewater discharge

Parameter	Unit	TULSMA limits	Vivero discharge	Marcopamba discharge	Negroyaco discharge	Vinchoa bridge discharge
COD	mg/L	200	543.08	465.28	210.67	740.06
BOD$_5$	mg/L	100	249.61	257.38	97.68	371.62
E. Coli	Col/L	2,000	2600	2,200	800	3,400
Total phosphorous	mg/L	10	12.70	12.83	9.87	16.47
Total iron	mg/L	10	11.01	14.75	10.02	15.68
Total manganese	mg/L	2	2.93	3.97	2.13	4.07
Total nitrogen	mg/L	50	61.67	52.97	50.69	55.46
Total suspended solids (TSS)	mg/L	130	120.07	135.34	110.46	172.39
Tensides	mg/L	0.5	3.64	2.60	1.97	8.94

4.3 Alternative choice

4.3.1 Alternatives proposal
The proposed alternatives in this research are:

1. Alternative 1: Design the interceptor that connects all the discharges with the existing plant, redesign the existing reactors and implement the necessary ones to carry out a purification with the following systems:
   a. Treatment line 1: Anaerobic, physical-chemical process and filter
   b. Treatment line 2: Percolating filter and clarifier
2. Alternative 2: Design two interceptors that connect the discharges in two groups, directing group 1 to a new plant and group 2 to the area of the existing plant using the following systems:
   a. Treatment line 1: Anaerobic, physical-chemical process and filter
   b. Treatment line 2: Percolating filter and clarifier

4.3.2 Likert comparative matrix
Table 3 shows the evaluation results in the Likert matrix of the alternatives proposed in 4.3.1. Once all the alternatives have been evaluated, the one with the highest score is selected as the most convenient.

With a result of 65 points, the selected alternative is A1b. This alternative corresponds to the interceptor's design that connects all discharges with the existing plant. Therefore, the present reactors were redesigned, and the necessary ones were implemented to carry out a purification with percolating filter and clarifier.

4.4 Interceptor-collector design

The interceptor-collector design consists of pipes plus four chambers placed in each of the identified discharges (see Fig. 1). As it is a single collector, which transports urban

Table 3: Selection of the optimal alternative using the Likert matrix.

	Alternative 1 (A1)		Alternative 2 (A2)	
	A1a	A1b	A2a	A2b
EMAPA-G preferences				
Design uses existing floor area	5	5	5	5
Design uses existing reactors	5	5	5	5
Economic Considerations				
No new land acquisition required	5	5	2	2
The design does not require earthwork	5	5	2	2
Topographical studies	4	4	3	3
Energy costs	3	3	2	2
Chemical consumption costs	2	5	2	5
Requires trained staff	3	3	3	3
Sludge management	2	3	2	3
Social considerations				
Pest presence	5	5	5	5
Generates bad odours	5	5	5	5
Technical considerations				
Contaminant removal above the required	5	4	5	4
Gravity designed	5	5	5	5
Environmental considerations				
Eutrophication risk in the receiving body or other aquifers	5	5	5	5
Damage to flora and fauna	3	3	2	2
TOTAL	62	**65**	53	56

wastewater and rainwater, the necessary dimensions must be able to control both flow types. Therefore, the interceptor directs only the wastewater and rainwater flow (10%) to the WTP. The rest of the rainwater flow must be diverted autonomously by the chambers, directly to the Guaranda river.

Four interceptor-collector chambers were designed, considering the following: (i) minimum capacity (allows personnel to enter for cleaning and maintenance), and (ii) speed and minimum diameter (time to avoid sedimentation). Table 4 shows the dimensions, flow to the interceptor-collector and the river, diameter, and velocity of the outlet pipe.

As an example of a design scheme, chamber 1 is observed in Fig. 5.

Table 4: Dimensions of the proposed chambers.

Parameters	Chamber 1	Chambers 2 & 3	Chamber 4
Volume (length × width × height) (m^3)	2.69 (1.8 × 1.8 × 0.83)	3.61 (1.9 × 1.9 × 1.0)	3.30 (1.8 × 1.8 × 1.02)
Outlet pipe diameter (m)	0.60	0.70	0.70
Output speed (m/s)	1.25	1.98	2.08
Input flow (m^3/s)	2.67	3.47	1.04
Output flow (m^3/s)	0.36	0.76	0.81
Flow to the river (m^3/s)	2.31	2.71	0.23

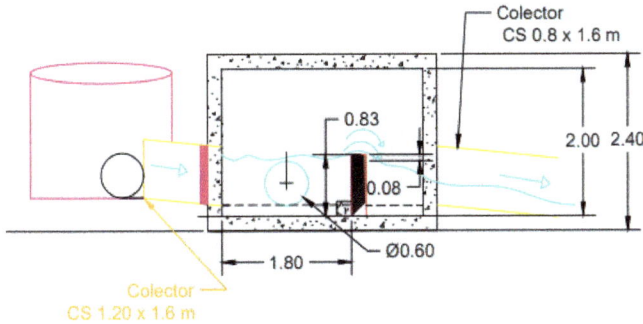

Figure 5: Scheme of interceptor-collector chamber 1.

4.5 WTP design

First, WTP regulation tank is designed. The regulation tank must store the flow derived by the chambers. Then, the stored water must be released to the WTP through an outlet pipe with a stop valve. The outlet pipe must be designed so that the flow rate does not exceed 20% of the WTP design flow rate. Therefore, to comply with this, the regulation tank sizing, the emptying pipe, and the pipe for sludge removal are carried out. Fig. 6 shows a scheme of the regulator tank.

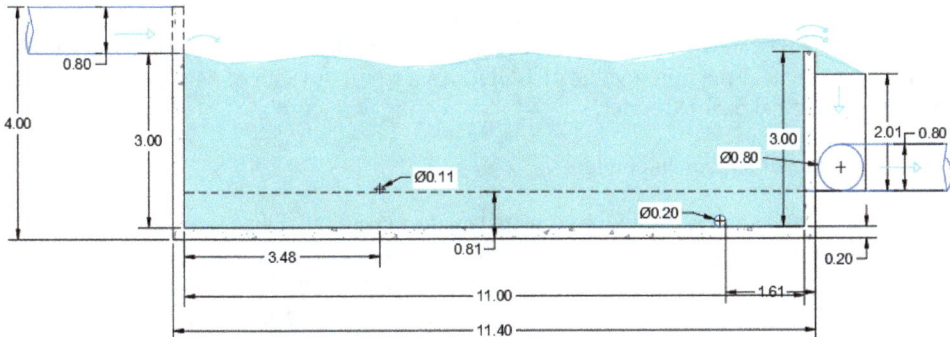

Figure 6: Cross-section of the regulator tank.

As a second part of the WTP, the preliminary treatment design is used to remove larger solids (plastic containers, bottles, branches, among others). For this, thick and thin bars were designed. For thick bars, 1 cm thick and with a separation of 2.50 cm between them are proposed. At the same time, fine bars are proposed with a thickness of 0.50 cm and a separation of 1.00 cm. After passing through the thick and fine bars, we proceed to the sand trap. The sand trap is a unitary operation that removes the sedimentable solids present in the raw wastewater. This is due to its length and hydraulic retention time.

The following system to design is a percolator filter. This filter is designed to obtain an effluent with a BOD_5 maximum of 80 mg/L. This level is below the BOD_5 maximum permissible limit of 100 mg/L for discharges to freshwater bodies given by the TULSMA. Furthermore, the contaminant load, which enters the filters (see Table 3), are results obtained in the EMAPA-G laboratory. Therefore, a percolating filter with a volume of 163.80 m^3 with a height of 1.50 m is proposed. Fig. 7 shows a scheme of the percolator filter proposed for the WTP.

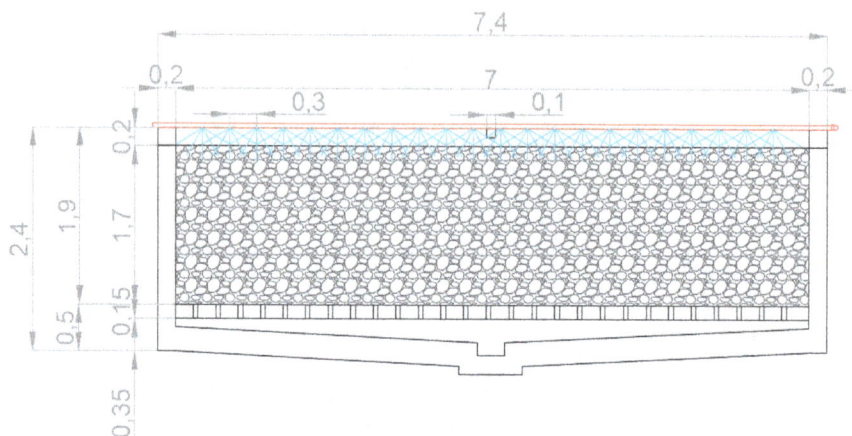

Figure 7: Cross-section of the percolator filter.

Lastly, a clarifier is required in WTP, prior to discharge to the receiving body to eliminate suspended sedimentable particles generated from the previous process.

With this WTP, it is possible to reduce the contaminants present in the wastewater, especially by 79% of the initial value of BOD$_5$. As a result, a value of 64 mg/L of BOD$_5$ is obtained in the final purified water.

4.6 Environmental impact and budget

The environmental aspects that will be affected by the execution of the project were identified through the analysis of the physical, biotic, and socioeconomic aspects, identifying that in each of them, there will be impacts in small proportions. The activities that are considered to have a high impact are: (i) the exploitation and transportation of imported borrow material (dust emission and soil contamination), and (ii) excavation and removal of material (soil contamination).

Lastly, an approximate evaluation of the budget of all the proposed designs has been made. The budget has been divided into two parts, one for the interceptor-collector chambers and another for the WTP. For the cameras, there is a budget of USD 554,946.61, while the WTP budget is USD 387,006.88. WTP cost is lower because old WTP tanks have been considered.

5 CONCLUSIONS

Guaranda, Ecuador, is a typical example of a city without a land-use plan that needs a wastewater treatment system. Alternatives with viable solutions for the problem were proposed considering the existing systems, restrictions, and favourable factors. The chosen design considers the transport and treatment of wastewater by the country's current regulations. The system promotes the health and well-being of the inhabitants and prevents Guaranda river contamination. Total removal levels achieved are less than 80 mg/L of BOD$_5$.

Alternative A1b was selected, which consists of designing an interceptor, four chambers and a Wastewater Treatment Plant (WTP) with two treatment lines. Each line is made up of the pre-treatment that consists of roughing and sand trap, a series of percolator filters and a clarifier. In addition, the use of the existing plant facilities is proposed.

The interceptor was designed to collect residual water and a percentage of rainwater (first washing water) through the sewage network towards the WTP. As a result, the purification system reduces the contaminants present in the wastewater by 79%, discharging the purified effluent to the receiving body with a BOD_5 value of 64 mg/L, complying with the discharge limits to a body of freshwater exposed in the TULSMA.

Environmental analysis and an estimated budget of the selected alternative were carried out. The transportation, exploitation and placement of loan material was the activity with the highest environmental impact. The budget, including the interceptor-collector and the WTP, reaches USD 941,953.49. For a design population of 32,922 inhabitants, the cost is USD 28.61 per inhabitant. This value is low because it should be considered that it is only about the interceptor and the purifying system; it does not include secondary, tertiary or home networks.

ACKNOWLEDGEMENTS

To the tutors and members of CIPAT-ESPOL who have collaborated in writing the research. To Edgar Berrezueta, Titular Scientist of the Geoscientific Infrastructure and Services Department (IGME-CSIC), for his help in reviewing the research. Also, to the reviewers and the editorial board for their comments and suggestions to improve the manuscript.

REFERENCES

[1] Qadir, M., Wichelns, D., Raschid-Sally, L., McCornick, P.G., Drechsel, P., Bahri, A. & Minhas, P.S., The challenges of wastewater irrigation in developing countries. *Agric. Water Manag.*, **97**, pp. 561–568, 2010.

[2] UNESCO, *WWAP Wastewater: The Untapped Resource*, Paris, France, 2017.

[3] Wang, D., Hubacek, K., Shan, Y., Gerbens-Leenes, W. & Liu, J., A review of water stress and water footprint accounting. *Water*, **13**, P. 201, 2021.

[4] Gómez-Limón, J.A., Gutiérrez-Martín, C. & Montilla-López, N.M., Agricultural water allocation under cyclical scarcity: The role of priority water rights. *Water*, **12**, p. 1835, 2020.

[5] Oliveira, G.A., Colares, G.S., Lutterbeck, C.A., Dell'Osbel, N., Machado, Ê.L. & Rodrigues, L.R., Floating treatment wetlands in domestic wastewater treatment as a decentralised sanitation alternative. *Sci. Total Environ.*, **773**, 145609, 2021.

[6] Merchan, B., Ullauri, P., Amaya, F., Dender, L., Carrión, P. & Berrezueta, E., Design of a sewage and wastewater treatment system for pollution mitigation in El Rosario, El Empalme, Ecuador. *WIT Transactions on Ecology and the Environment*, vol. 251, WIT Press: Southampton and Boston, pp. 77–85, 2021.

[7] Moreira, F.D. & Dias, E.H.O., Constructed wetlands applied in rural sanitation: A review. *Environ. Res.*, **190**, 110016, 2020.

[8] Ramos, N. de F.S., Borges, A.C., Coimbra, E.C.L., Gonçalves, G.C., Colares, A.P.F. & de Matos, A.T., Swine wastewater treatment in constructed wetland systems: Hydraulic and kinetic modelling. *Water*, **14**, p. 681, 2022.

[9] Carballo, F.M., Brito, L.M., Mero, P.C., Aguilar, M.A. & Ramírez, J.T., Urban wastewater treatment through a system of green filters in the Montañita commune, Santa Elena, Ecuador. *WIT Transactions on Ecology and the Environment*, vol. 239, WIT Press: Southampton and Boston, pp. 233–249, 2019.

[10] Rosari, N.L. & Purwanti, I.F., Design of sewerage system and wastewater treatment plant in Asemrowo, Surabaya, Indonesia. *IOP Conf. Ser. Earth Environ. Sci.*, **506**, 012021, 2020.

[11] Merchán-Sanmartín, B., Aguilar-Aguilar, M., Morante-Carballo, F., Carrión-Mero, P., Guambaña-Palma, J., Mestanza-Solano, D. & Berrezueta, E., Design of sewerage system and wastewater treatment in a rural sector: A case study. *Int. J. Sustain. Dev. Plan.*, **17**, pp. 51–61, 2022.

[12] Singh, N.K., Kazmi, A.A. & Starkl, M., A review on full-scale decentralised wastewater treatment systems: Techno-economical approach. *Water Sci. Technol.*, **71**, pp. 468–478, 2015.

[13] Massoud, M.A., Tarhini, A. & Nasr, J.A., Decentralised approaches to wastewater treatment and management: Applicability in developing countries. *J. Environ. Manage.*, **90**, pp. 652–659, 2009.

[14] Pozo, M., Serrano, J.C., Castillo, R. & Moreno, L., Indicadores de Agua, Saneamiento e Higiene en Ecuador, Quito, Ecuador, 2016.

[15] Herrera-Franco, G., Alvarado-Macancela, N., Gavín-Quinchuela, T. & Carrión-Mero, P., Participatory socio-ecological system: Manglaralto-Santa Elena, Ecuador. *Geology, Ecology, and Landscapes*, **2**(4), pp. 303–310, 2018.

[16] González, R. & Vera, H., Diseño del interceptor y sistema depurador para las descargas directas de aguas residuales al río Guaranda, provenientes de la red de alcantarillado combinado de las parroquias urbanas Veintimilla y Polibio Chávez de la ciudad de Guaranda, ESPOL Polytechnic University, 2022.

[17] INEC, Resultados del censo de población y vivienda de 2010 en Ecuador. https://www.ecuadorencifras.gob.ec/censo-de-poblacion-y-vivienda/. Accessed on: 4 Apr. 2022.

[18] INAMHI-Ecuador, Anuario Meteorológico 2013. http://www.serviciometeorologico.gob.ec/docum_institucion/anuarios/meteorologicos/Am_2013.pdf*. Accessed on: 3 Apr. 2022.

[19] SNI-Ecuador, Archivos de Información Geográfica. https://sni.gob.ec/coberturas. Accessed on: 27 Mar. 2022.

[20] CPE-INEN, Código De Práctica Ecuatoriano: Normas para el diseño de agua potable y disposición de aguas residuales para poblaciones mayores a 1000 habitantes, Quito, Ecuador, 1992.

[21] Buroz, E., Métodos de Evaluación de Impactos, La Plata, Uruguay, 1994.

[22] GAD de Guaranda, Población de la ciudad de Guaranda. http://www.guaranda.gob.ec/newsiteCMT/datos-importantes/. Accessed on: 26 Apr. 2022.

EVOLUTION OF THE ACTIVATED SLUDGE COMMUNITY OF A WASTEWATER TREATMENT PLANT WITH INDUSTRIAL DISCHARGES

ÁNGELA BAEZA-SERRANO[1], FELIU SEMPERE[1], NURIA OLIVER[1],
PILAR GUTIÉRREZ[2] & GLORIA FAYOS[1]
[1]Global Omnium Medioambiente, S.L., Spain
[2]Explotaciones Hidricas del Cinca, SA, Spain

ABSTRACT
Stability and high microbial activity of the biological processes in wastewater treatment plants (WWTPs) based on activated sludge processes, is a key factor for the correct treatment of wastewater prior to its discharge into the environment. This can be a challenge, as WWTPs are obliged to treat all the wastewater they receive, including possible uncontrolled discharges, some containing toxic substances or poorly balanced loads to the microorganisms in charge of carrying out the wastewater treatment. In fact, the effects of these pollutants on biological treatment by activated sludge include inhibition of bacteria in the removal of organic compounds and nutrients, reduction of the efficiency of solids separation and modification of the compaction properties of the sludge, effects that can have a negative impact on the treatment processes and on the quality of WWTP effluents. The aim of this study is to describe the composition and changes of the bacterial community in a Spanish full-scale activated sludge WWTP which suffers seasonal discharges from the vegetable canning industry during a year.
Keywords: activated sludge, metagenomic analysis, industrial discharges, bacterial community.

1 INTRODUCTION

There are more than 30,000 urban wastewater treatment plants (WWTPs) in the EU [1], including 80% of them secondary treatment using activated sludge. In countries from south east Europe, such as Romania and Bulgaria, this percentage is much lower, due to the problems associated to maintaining the biological process. The activated sludge is a biological process consisting in a bacterial culture growing in an aerated reactor. This culture is able to metabolize organic compounds, nutrients (nitrogen and phosphorus) and other substances present in wastewater at given design loads. The produced sludge sediments and approximately 90% is recirculated back to the bioreactor to maintain the process biomass. The remaining 10% is separated in a settler and removed from the system to maintain the solids concentration constant.

Even though there are important economic investments for the creation and renovation of urban WWTP [2], the increase of the urban wastewater toxic load hinders the treatment. During the year, the entry of spills greater the load which cannot be eliminated by the bacterial culture. Uncontrolled discharges containing harmful substances or high loads can damage the biological units, reducing the cleaning capacity of the bacterial culture or creating nutritional imbalances that favors the growth of filamentous morphotypes that cause sludge separation problems such as bulking or foaming. These discharges are often seasonal and industrials coincide with campaigns such as the pouring of wine or canning companies.

In the vegetable canning sector, approximately 70%–80% of the water consumption is discharged as wastewater and the remaining 20%–30% is incorporated into the product or lost in evaporation. In most cases, work is done on a seasonal basis, taking advantage of the availability of different raw materials throughout the year. In each season, different vegetables are used, sometimes requiring different manufacturing stages and therefore producing a significant change in the levels of water consumption as well as in the

WIT Transactions on Ecology and the Environment, Vol 259, © 2022 WIT Press
www.witpress.com, ISSN 1743-3541 (on-line)
doi:10.2495/AWP220121

characteristics of the wastewater generated. Regarding the pollutant load of these discharges, it is basically composed of organic matter (biochemical oxygen demand at 5 days, BOD5 and chemical oxygen demand, COD) and suspended solids (SS), and sometimes there are also discharges with high conductivity and variable pHs depending on the cleaning processes or if alkaline peeling is used. Contamination levels increase significantly in operations such as blanching and cleaning of facilities. In blanching, the water is loaded with organic matter, COD, due to the dissolution of substances such as sugars, starches and soluble organic products from the vegetables. Depending on the raw material used, these are the levels of contamination that can be found in the discharge are between 600–12,000 mg/l COD and 100–3,000 mg/l SS [3].

The wastewater from the municipality of Binaced (Fig. 1) is a double-stage system treating 1,445 p.e. with an average flow of 600 m^3/day. Wastewater is treated in a 190 m^3 integrated fixed-film activated sludge (IFAS) system equipped with a reagent dosing system. The water then passes into the biological reactor with a volume of 711 m^3, from where it is distributed to the secondary settling tanks.

Figure 1: Binaced WWTP (Huesca, Spain).

A percentage of the sludge extracted from the secondary settling tanks is recirculated to the IFAS system, while the remaining portion is pumped to the thickeners, from where it is finally dewatered and destined for agricultural application.

The influent received by the Binaced WWTP is largely conditioned by the campaign of the vegetable cannery industry in the municipality, since during the period July–December both the organic load and the flow received by the facility increase. These industrial inputs vary depending on the year and the evolution of the harvests, but they tend to follow a similar pattern every year: July–September canned peach, September–October canned fruit cocktail and November–December canned pear.

This kind of discharges are characterized by a high COD load, causing a nutritional imbalance because sufficient nutrients are required to degrade organic matter and satisfy bacterial requirements both for growth and good floc formation. Moreover, nutrient limitation has been linked to the filamentous microorganisms growth causing poor settling [4].

Binaced WWTP influent commonly presents imbalances between organic matter and nutrients during these incidents, with the addition of urea and/or phosphoric acid being necessary to achieve the desirable ratio of degradable organic matter to available nutrients, BOD:N:P ratio, of 100:5:1 [5].

In this paper, a study of the discharges effects in mixed liquor microbiome of biological reactor of Binaced WWTP is presented. Full-scale studies are important since the selective pressures are likely very different compared to in small-scale systems.

2 MATERIALS AND METHODS

Monthly samples of 5 ml were taken in three different points of biological reactor and were immediately frozen for their transport on dry ice to ADM-Lifesequencing S.L. laboratory (Paterna, Valencia, Spain). During the month of April, it was not possible to take samples because the biological reactor had to be emptied for maintenance work.

Genomic DNA was extracted from 2 mL of the mixed activated sludge samples with Qiamp Power Fecal Mini kit (Qiagen) with enzymatic lysis and mechanic disruption. DNA were amplified following the 16S Metagenomic Sequencing Library Illumina 15044223 B protocol (Illumina). In brief, the first amplification step, primers were designed containing: a universal linker sequence allowing amplicons for incorporation indexes and sequencing primers by Nextera XT Index kit (Illumina); and 16S rRNA gene universal primers [6] and in the second and last amplification indexes were included. Libraries were quantified by fluorimetry using Quant-iT™ PicoGreen™ dsDNA Assay Kit (Thermofisher) and pooled before to sequencing on the MiSeq platform (Illumina), configuration 300 cycles paired reads. The size and quantity of the pool were assessed on the Bioanalyzer 2100 (Agilent) and with the Library Quantification Kit for Illumina (Kapa Biosciences), respectively. PhiX Control library (v3) (Illumina) was combined with the amplicon library (expected at 20%). Sequencing data were available within approximately 56 hours. Image analysis, base calling and data quality assessment were performed on the MiSeq instrument.

For massive sequencing, the hypervariable region V3–V4 of the bacterial 16s rRNA gene was amplified using key-tagged eubacterial primers [7] and sequenced with a MiSeq Illumina Platform, following the Illumina recommendations.

The resulting sequences were split taking into account the barcode introduced during the PCR reaction, while R1 and R2 reads were overlapped using PEAR program version 0.9.1 [8] providing a single FASTQ file for each of the samples. Quality control of the sequences was performed in different steps, (i) quality filtering (with a minimum threshold of Q20) was performed using fastx tool kit version 0.013, (ii) primer (16s rRNA primers) trimming and length selection (reads over 300 nts) was done with cutadapt version 1.4.1 [9]. These FASTQ files were converted to FASTA files and UCHIME program version 7.0.1001 was used in order to remove chimeras that could arise during the amplification and sequencing step. Those clean FASTA files were BLAST [10] against NCBI 16s rRNA database using blastn version 2.2.29+. The resulting XML files were processed using a python script developed by

ADM-Lifesequencing S.L. (Paterna, Valencia, Spain) in order to annotate each sequence at different phylogenetic levels (Phylum, Family, Genus and Species).

3 RESULTS AND DISCUSSION

3.1 Bacterial community structure

In total, 215,501 effective sequences of the 16S rRNA gene were generated from 11 samples, similar that those found by Hu et al. [11] (202,968 in 16 samples) that widely represented the diversity of the microbial communities.

The calculated species richness Chao1 and Shannon index are shown in Table 1. Gonzalez-Martínez et al. [12] studied 10 different wastewater treatment systems from Spain and Netherlands and found Chao 1 and Shannon index between 1395,003 and 441,150 and 5.137 and 2.831 respectively, lower than those found in our study. High Shannon index values obtained point that activated sludge ecosystems are very diverse.

Table 1: Chao 1 Richness Estimation and diversity index.

Month	Shannon value	CHAO1
February	5,429	2012
March	5,366	2001
May	5,321	2647
June	5,435	2497
July	5,299	2104
August	4,989	2641
September	5,145	2724
October	5,378	2542
November	5,135	2465
December	5,443	2967
January	5,466	2886

Fig. 2 shows the heat map of core genera relative abundance in activated sludge samples during a year. Taxa represented occurred in a great than 1% relative abundance in at least one sample. Between 4.79% and 15.39% of the sequencing reads could not be assigned to any taxa at genera level. These "no hit" are lower values comparing with the 30% found by Wang et al. [13] or the 32%–34% described in Zhang et al. [14].

OTU genera with relative abundance in less than 1% were between 59.57% and 47.56%.

Significant values of the denitrifying Flavobacterium were found at the first half of the year. Flavobacterium was previously described as a part of community core in sewage plants and reported as floc-forming microorganisms [15]. Zhang et al. [14] also reported Flavobacterium as a dominant genus in three WWTP from North America of the 14 samples studied. However, a drop in its abundance is observed coinciding with the start of the peach canning season in July, and with discharges with high pH. A Similar drop is observed for Thermomonas, which has been described as part of core members of the denitrifying communities in full-scale WWTPs too [16], and the denitrifying Rhodoferax and Ferruginibacter, coinciding also with a decrease in COD elimination efficiency detected after a high Conductivity discharge on 18 July (Fig. 3).

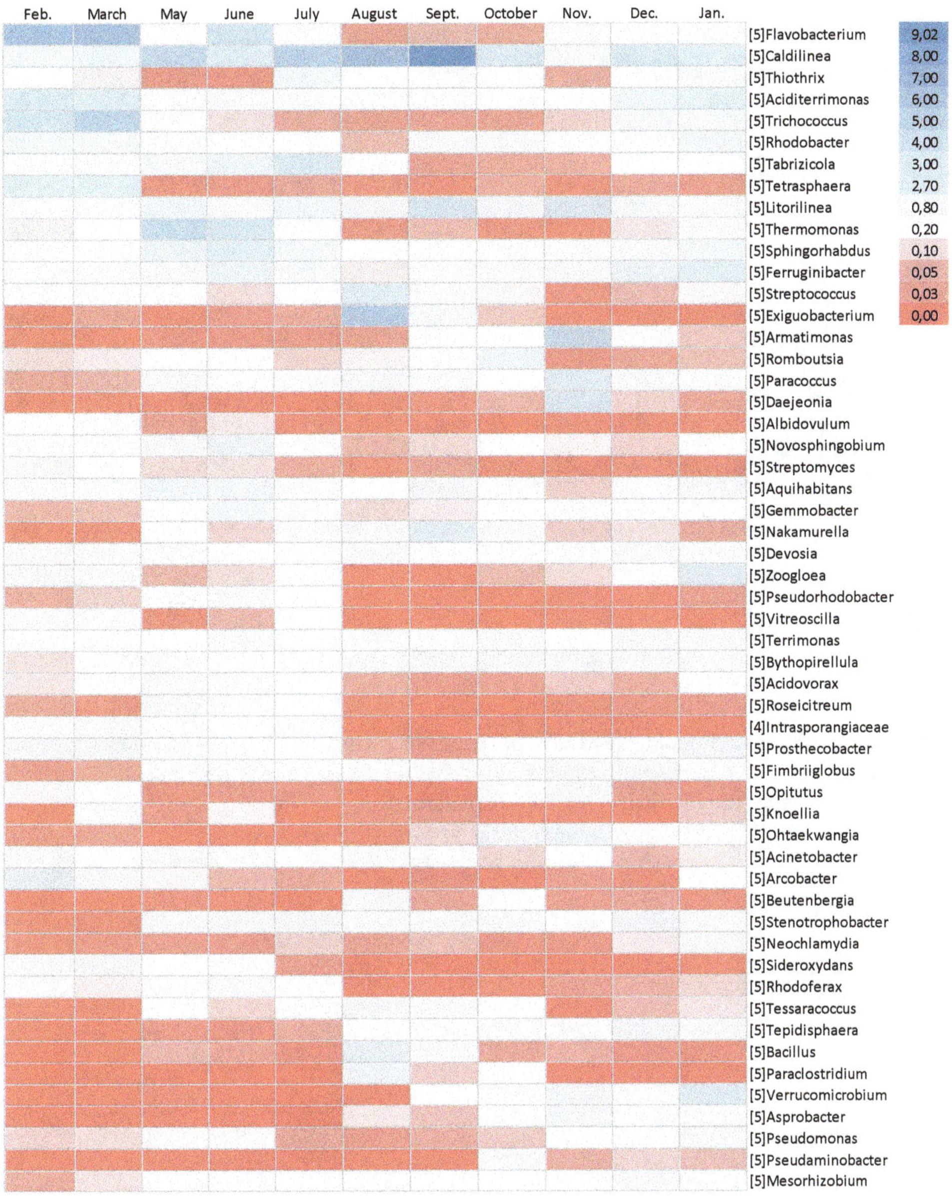

Figure 2:   Heat map of abundance of genera OTUs > 1% in at least one sample, presented in terms of percentage from the total number of bacterial sequences in each sample. Confidence threshold of taxonomical identification was 99%.

On the other hand, the relative abundance of Exiguobacterium genus increased in August, after the start of the campaign mentioned above. Exiguobacterium has been described as thermotolerant, also, capable of growing within a wide range of pH values (5–11) and

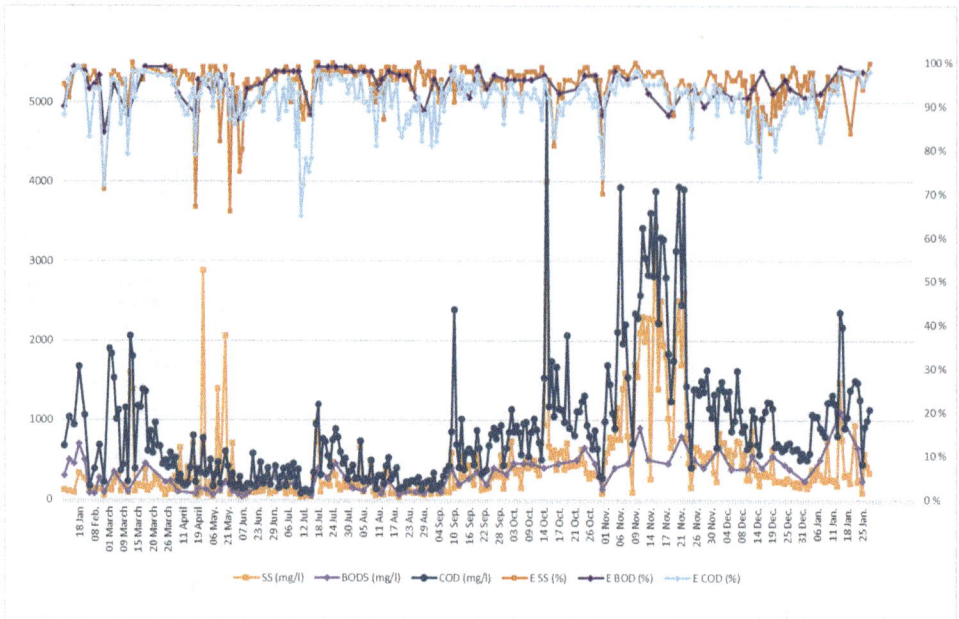

Figure 3: Evolution of organic matter and SS input and removal efficiency.

alkaliphilic, tolerant to high levels of UV radiation, and heavy metal stress (including arsenic) [17]. Some genera like Stenotrophobacter (represented by the only specie *S. terrae*), Fimbriiglobus and the denitrifying Tepidisphaera seem to be increasing their relative abundance coinciding with rising temperatures in May (Fig. 4).

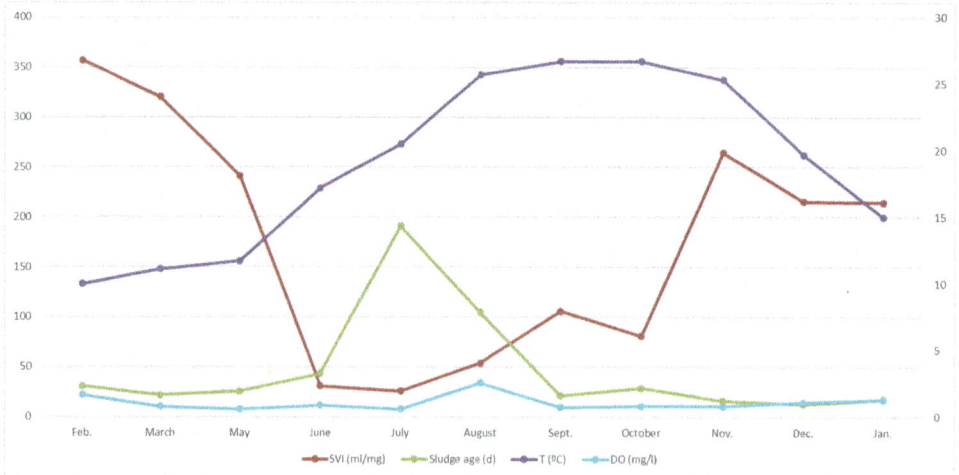

Figure 4: Evolution of temperature, sludge volume index, sludge age and dissolved oxygen in biological reactor.

With increasing COD in November (Fig. 3), fermentative acidogenic bacteria such as Opitutus, which is associated with the use of moderately recalcitrant heteropolysaccharides such as xylan and pectin [18], both present in fruit, also increase. Armatimonas, on the other hand, also increased its relative abundance. They have been described in carbon removing plants, typically of high load with a large input of industrial wastewater. They grow on a limited range of substrates, such as sucrose, raffinose, gentiobiose (present in peach stones) and pectin [19].

According to Wang et al. [13] there were 60 genera of bacterial populations commonly shared by all samples, including Ferruginibacter, Prosthecobacter, Zoogloea, Subdivision 3 genera incertae sedis, Gp4, and Gp6, indicating that there is a core microbial community in the microbial populations of WWTPs at different geographic locations. Both, Ferruginibacter and Zoogloea, as well as Prosthecobacter were found at abundances greater than 1% (Fig. 2).

3.2 Evolution of nitrifying genera

High biodiversity of nitrifying bacteria was detected. Ten species of Ammonia Oxidizing Bacterium (AOB) were found, including the genera Nitrosomonas, Nitrosococcus and Nitrosospira, being *Nitrosomonas oligotropha* the most abundant (with relative abundances between 0.641% and 0.039%). Regarding Nitrite oxidizing bacteria (NOB), seven species were detected, *Nitrospira moscoviensis* was the most abundant (0.348%–0.007%), but also representatives of Nitrobacter, Nitrolancea and nitrite reductor Rhizobium [13] were found.

AOB bacteria represented between 0.200% and 0.006% of samples relative abundance and NOB between 0.523% and 0.122%. Zhang et al. [14] indicated that AOB occupied 0%–0.64% of the total pyrosequencing sequences from six municipal WWTPs, and Nitrosomonas dominated in the WWTPs. Kim et. al. [20] described similar NOB amount (0.06% of Nitrospira-like nitrite-oxidizing bacteria) in suspended activated sludge of an IFAS system, and suggested that the suspended biomass contributed more to nitrification than did the attached biomass in a IFAS system.

Fig. 5 shows nitrifying bacteria evolution during the year. An increase in relative abundance is observed in July, with a subsequent decline in abundance, interestingly coinciding with the nitrogen removal efficiency loss (Fig. 6). The increase of abundance in May and June could be related to higher temperatures and sludge ages (Fig. 4).

3.3 Evolution of microorganisms causing bulking and/or foaming

High SVI were detected during February, March and again in November (Fig. 4). In the first months, two genera related with Nostocoida limicola-like filamentous morphotype were found: Trichococcus and Tetrasphaera, both related with low temperatures [19].

Caldilinea is a gram-negative and facultatively aerobic bacteria which is able to grow both aerobically and anaerobically (fermentatively), and is phenotypically recognized as Eikelboom's morphotype 0803 in activated sludge [21]. This genus was detected at high relative abundances from May to October, probably related to high temperatures (Fig. 4), and then increase again in December, maybe influenced by the numerous discharges recorded during November.

Also during November, the biodegradability of the sewage decreased by the presence of pear cannery discharges. This could be related with the drop of bulking forming bacteria Thiothrix, which is described as a consumer of rapidly biodegradable substrates such as volatile fatty acids.

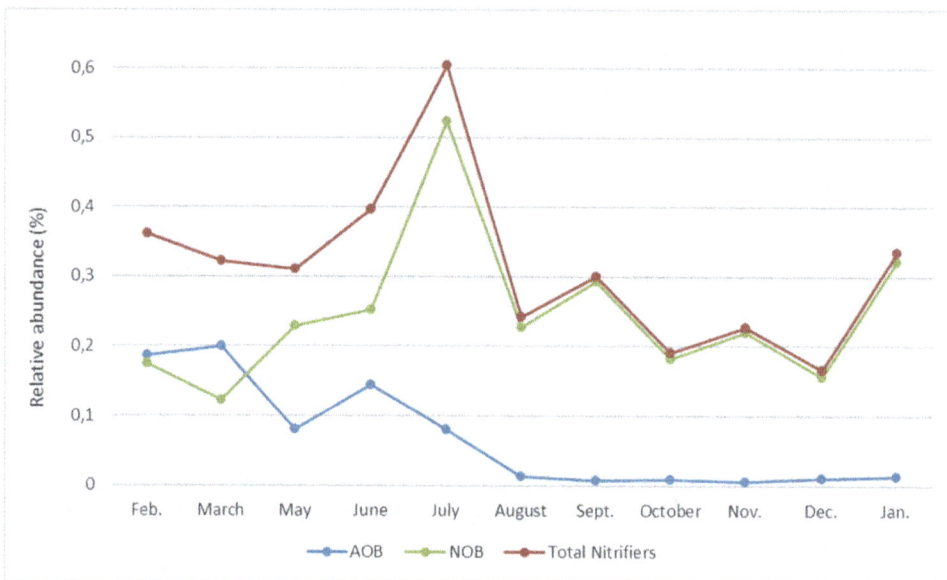

Figure 5: Abundance of nitrifying genera OTUs presented in terms of percentage from the total number of bacterial sequences in each sample. Confidence threshold of taxonomical identification was 99%.

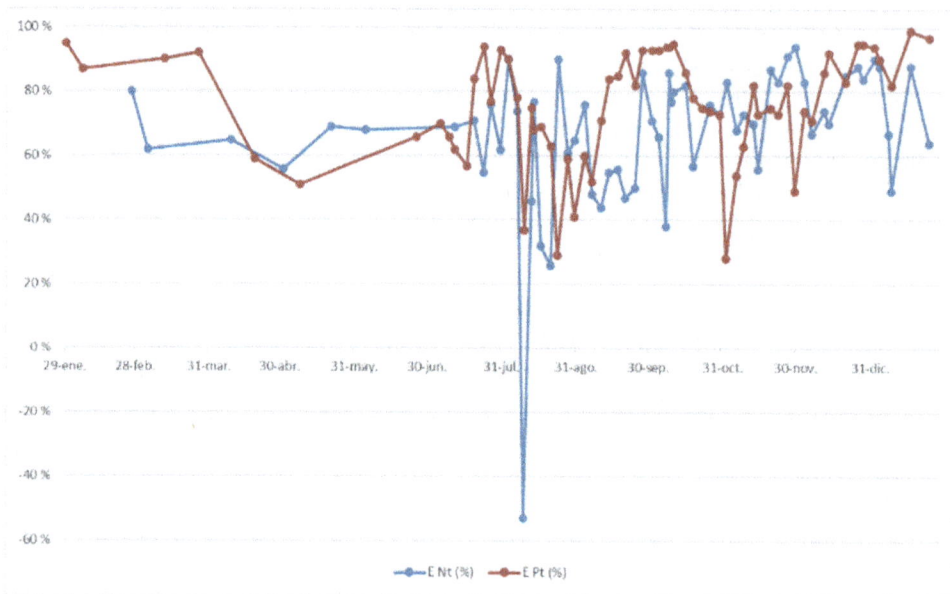

Figure 6: Nitrogen and phosphorus removal efficiency.

4 CONCLUSIONS

This study describes the composition and changes of the bacterial community in a full-scale activated sludge WWTP suffering seasonal discharges from the vegetable canning industry during a year.

Fifty-four genera were found in high relative abundance at least one sample, accounting between 40% and 27% of total genera.

Certain denitrifying genera like Flavobacterium, Thermomonas, Rhodoferax and Ferruginibacter, decreased coinciding with the start of the peach canning season and with discharges with high pH.

Certain bacteria related to the use of sugars present in the fruit and its stones, such as Opitutus and Armatimonas, significantly increase their relative abundance during the canning seasons.

On the other hand, the relative abundance of the alkaliphilic genera Exiguobacterium increased after the start of the peach cannery campaign.

Some genera like Stenotrophobacter and the denitrifying Tepidisphaera increased their relative abundance coinciding with rising temperatures.

Nitrifying bacteria do not seem to be affected by fruit cannery campaigns but are affected by discharges with high pH and conductivity.

ACKNOWLEDGEMENTS

The authors would like to thank the funding from the European Commission through LIFE Programme (LIFE ENV16/ES/000390). The opinions or points of view published herein do not represent EC official position. The authors would also like to thank the Instituto Aragonés del Agua, attached to the Department of Agriculture, Livestock and Environment of the Government of Aragón (Spain).

REFERENCES

[1] European Environment Agency, Database. https://sdi.eea.europa.eu/catalogue/srv/eng/catalog.search#/metadata/38ee29b5-70b1-4431-abf6-7c1aa3c50515. Accessed on: 1 Feb. 2022.

[2] Report from the Commission to the European Parliament, the Council and the European Economic and Social Committee and the Committee of the Regions. Eighth Report on the Implementation Status and the Programmes for the Implementation (as required by Article 17) of Council Directive 91/271/ECC concerning urban waste water treatment.

[3] Aguas Industriales. http://aguasindustriales.es/aguas-residuales-en-el-sector-de-las-conservas-vegetales/. Accessed on: 1 Feb. 2022.

[4] Jenkins, D., Richard, M.G. & Daigger, G.T., *Manual on the Causes and Control of Activated Sludge Bulking, Foaming and Other Solids Separation Problems*, CRC Press, 2004.

[5] Henze, M., Harremoes, P., Arvin, E. & La Cour Jansen, J., Wastewater treatment. *Biological and Chemical Processes*, 2nd ed., Germany, 1997.

[6] Klindworth, A., Pruesse, E., Schweer, T., Peplies, J., Quast, C., Horn, M. & Glöckner, F.O., Evaluation of general 16S ribosomal RNA gene PCR primers for classical and next-generation sequencing-based diversity studies. *Nucleic Acids Research*, **41**(1), p. e1, 2013.

[7] Zhang, J., Kobert, K., Flouri, T. & Stamatakis, A., PEAR: A fast and accurate Illumina Paired-End reAd mergeR. *Bioinformatics*, **30**(5), pp. 614–620, 2014.

[8]   Martin, M., Cutadapt removes adapter sequences from high-throughput sequencing reads. *EMBnet.journal*, **17**, pp. 10–12, 2011.

[9]   Altschul, S.F., Gish, W., Miller, W., Myers, E.W. & Lipman, D.J., Basic local alignment search tool. *Journal of Molecular Biology*, **215**, pp. 403–410, 1990.

[10]   Meerbergen, K., Van Geel, M., Waud, M., Willems, K.A., Dewil, R., Van Impe, J., Appels, L. & Lievens, B., Assessing the composition of microbial communities in textile wastewater treatment plants in comparison with municipal wastewater treatment plants. *MicrobiologyOpen*, **6**(1), 2017. DOI: 10.1002/mbo3.413.

[11]   Hu, M., Wang, X., Wen, X. & Xia, Y., Microbial community structures in different wastewater treatment plants as revealed by 454-pyrosequencing analysis. *Bioresource Technology*, **117**, pp. 72–79, 2012.

[12]   Gonzalez-Martínez, A., Rodriguez-Sanchez, A., Lotti, T., Garcia-Ruiz, M.-J., Osorio, F., Gonzalez-Lopez, J. & van Loosdrecht, M.C.M., Comparison of bacterial communities of conventional and A-stage activated sludge systems. *Sci. Rep.*, **6**, 18786, 2016. DOI: 10.1038/srep18786.

[13]   Wang, Z., Zhang, X.-X., Lu, X., Liu, B., Li, Y., Long, C. & Li, A., Abundance and diversity of bacterial nitrifiers and denitrifiers and their functional genes in tannery wastewater treatment plants revealed by high-throughput sequencing. *PLoS One*, **9**(11), e113603, 2014. DOI: 10.1371/journal.pone. 0113603.

[14]   Zhang, T., Shao, M.F. & Ye, L., 454 pyrosequencing reveals bacterial diversity of activated sludge from 14 sewage treatment plants. *ISME J.*, **6**(6), pp. 1137–1147, 2012.

[15]   Guo, F., Zhang, S.-H., Yu, X. & Wei, B., Variations of both bacterial community and extracellular polymers: The inducements of increase of cell hydrophobicity from biofloc to aerobic granule sludge. *Bioresour. Technol.*, **102**, pp. 6421–6428, 2011.

[16]   McIlroy, S.J., Starnawska, A., Starnawski, P., Saunders, A.M., Nierychlo, M., Nielsen, P.-H. & Nielsen, J.P., Identification of active denitrifiers in full-scale nutrient removal wastewater treatment systems. *Environmental Microbiology*, **18**(1), pp. 50–64, 2016.

[17]   Rodrigues J.L.M. & Isanapong J., The Family Opitutaceae. *The Prokaryotes*. Eds E. Rosenberg, E.F. DeLong, S. Lory, E. Stackebrandt & F. Thompson, Springer: Berlin, Heidelberg, 2014. DOI: 10.1007/978-3-642-38954-2_147.

[18]   Ordoñez, O.F., Lanzarotti, E., Kurth, D., Gorriti, M.F., Revale, S., Cortez, N., Vazquez, M.P., Farías, M.E. & Turjanski, A.G., Draft genome sequence of the polyextremophilic exiguobacterium sp. strain S17, isolated from Hyperarsenic Lakes in the Argentinian Puna. *Genome Announcements*, **1**(4), e00480-13, 2013. DOI: 10.1128/genomeA.00480-13.

[19]   Dueholm, M.S., Nierychlo, M., Andersen, K.S., Rudkjøbing, V., Knudsen, S., The MiDAS Global Consortium, Albertsen, M., & Nielsen, P.H., MiDAS 4: A global catalogue of full-length 16S rRNA gene sequences and taxonomy for studies of bacterial communities in wastewater treatment plants, 2021. DOI: 10.1101/2021.07.06.451231.

[20]   Kim, T.S., Kim, H.S., Kwon, S. & Park, H.D., Nitrifying bacterial community structure of a full-scale integrated fixed-film activated sludge process as investigated by pyrosequencing. *J. Microbiol. Biotechnol.*, **21**(3), pp. 293–298, 2011.

[21]   Kragelund, C., Thomsen, T.R., Mielczarek, A.T. & Nielsen, P.H., Eikelboom's morphotype 0803 in activated sludge belongs to the genus Caldilinea in the phylum Chloroflexi. *FEMS Microbiology Ecology*, **76**(3), pp. 451–462, 2011. DOI: 10.1111/j.1574-6941.2011.01065.x.

HOLISTIC APPROACH TO THE ECONOMIC BENEFITS OF USING RECLAIMED WATER IN AGRICULTURE

MARÍA JOSÉ LÓPEZ SERRANO*
Department of Business and Economics, University of Almería, Spain

ABSTRACT

Implementing water alternatives to ensure its supply is crucial in a worldwide context where water scarcity is a daily problem that is anticipated to intensify in the upcoming years. In this context, reclaimed water has proved to be a viable option for ensuring water availability for such a water-demanding and critical sector as agriculture is. Nevertheless, its use is too often seen as a last-resource alternative desired mainly by farmers facing water scarcity or a lack of water alternatives. Despite the pressing need to look for reasons that justify the profitability of its implementation over other water alternatives, and in light of there being gap research when analyzing all the benefits of using reclaimed water from a holistic economic approach, the overall objective of this study is to evaluate the economic impact generated by the use of tertiary water in agriculture to irrigate crops. To reach this ambitious objective, the economic value of a large list of positive externalities derived from the implementation of reclaimed water in agriculture is going to be calculated. Some of the aspects that are going to be evaluated are: its fertilizer value affecting crop productivity, soil quality and plant growth and its capacity of being a source of organic carbon. Furthermore, these others relevant aspects that too often go unnoticed due to the difficulty in their measurement are going to be evaluated: its contribution to protect water bodies and promote water resources quality and recovery, its capacity of enhancing and boosting marine ecosystem services and its leverage effect in promoting agriculture in new areas and in fostering different crop varieties in other areas where, because of salinity levels or water scarcity, only some crop varieties may be grown. Results will show through economic data and results the urge of implementing reclaimed water in agriculture worldwide.

Keywords: water reuse, reclaimed water, sustainability, circular economy, agricultural water contamination, wastewater management.

1 INTRODUCTION

Even distribution of a finite resource like water is today a serious concern, given that over 4 billion people experience severe water scarcity for at least a month each year [1], [2]. In light of this, agriculture's importance should be highlighted, as it is the world's largest water consumer, consuming up to 90% of total water supplies in some areas and predicted to worsen in many others [3]–[7]. Agriculture is also critical in terms of climate change, which is one of the greatest threats to both humans and the environment, because, in addition to its high vulnerability due to extreme and unpredictable weather conditions that exacerbate water scarcity, agriculture contributes up to 14% of global greenhouse gas emissions. [8]–[12]. Considering this, and because of the natural relationship between water and land use, agriculture plays a critical part in the worldwide job of ensuring sustainable water supply [13]–[16]. The quest for the implementation of a successful, long-lasting, and sustainable water source to irrigate crops is critical to agriculture satisfying the need of an ever-growing population [17]. This irrigation option should also be capable of ensuring economic, environmental, and social development [18]. To combat this alarming scenario, tertiary water use has been bolstered in recent decades, and even more so in recent years, as it is a resource that, in addition to being apparently inexpensive, helps to preserve important water supplies when utilized for irrigation [19], [20]. Another aspect of using treated wastewater to irrigate

---

*ORCID: https://orcid.org/0000-0001-6439-0819

WIT Transactions on Ecology and the Environment, Vol 259, © 2022 WIT Press
www.witpress.com, ISSN 1743-3541 (on-line)
doi:10.2495/AWP220131

crops in agriculture that merits special attention due to its importance and potential is its contribution to the circular economy (CE), as the treatment process facilitates the transformation of wastewater from waste to resource through reuse and recycling. [21], [22]. Even though it is becoming more frequently used due to its well-documented environmental, social, and economic benefits, reclaimed water irrigation is still at a low level due to a lack of social acceptance [23]–[25].

2 AREA OF STUDY

Because of its representativeness, the current study was conducted in the province of Almeria, which is a Mediterranean region in southeastern Spain. This area, shown in Fig. 1, is a semi-arid terrain with significant solar radiation due to its low annual rainfall of less than 200m and a mean annual temperature of 18°C [26]. Furthermore, irrigation water availability influences productivity and is one of the key limiting factors in one of Europe's driest regions [27], [28]. This region is well-known for its important agricultural role, having the world's highest concentration of greenhouses and being the largest exporter of fruits and vegetables in the European Union [29]–[32]. Due to localized irrigation, automated fertigation, and the use of tensiometers, the system used to irrigate crops in the region has undergone a significant upgrade in recent years, all of which have improved the efficiency of the water in the area [30]. Irrigated regions are split geographically into irrigation villages, each of which has its own water source. This semi-arid region needs special attention since its case study can be applied to other countries in comparable situations, particularly those with arid and semi-arid climates and water resource constraints [33]–[37]. The province is divided into six primary sub-areas, two of which hold more than 77% of the province's greenhouse gas concentration [38]. These two areas, Campo de Dalias and Bajo Andarax (which includes Almeria City), are crucial to this research because in Campo de Dalias tertiary water is not used for irrigation because the main water resource is groundwater, whereas in Bajo Andarax, treated

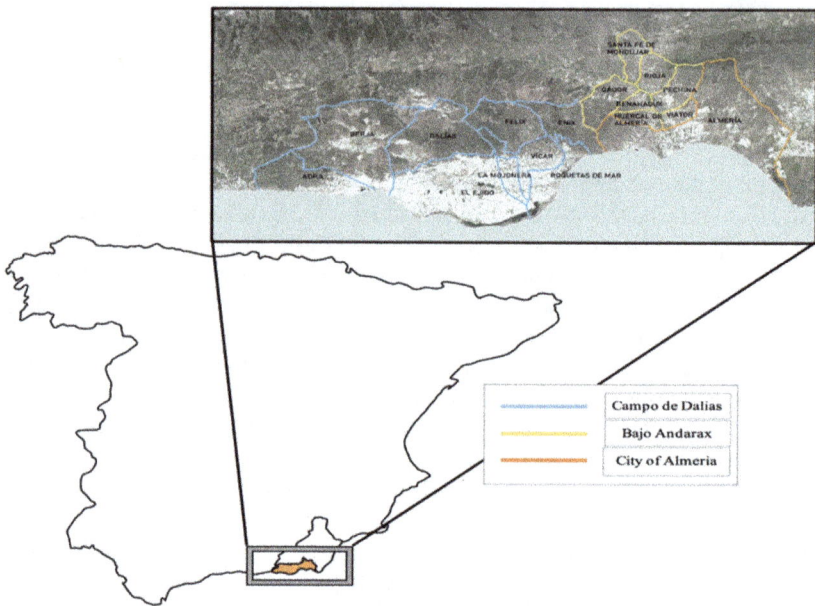

Figure 1: Area of study.

wastewater has been used for irrigation since 1997 [39]. These singularities make the area ideal for a comparison of the consequences of using reclaimed water for more than 20 years with others where it has never been used before.

After previous analyses in many geographical points, two farm plots with very similar soil structure, nearly identical level of precipitation and average temperatures have been chosen for this research. These two chosen farm plots are not adjacent because water filtration of both water sources may bias the results of the aspects analyzed.

3 METHODOLOGY

All the benefits of using reclaimed water are going to be analyzed from a holistic economic approach with the main objective of evaluating the direct and indirect economic impact generated by the use of tertiary water in agriculture to irrigate crops.

The following aspects are going to be measured through different methodologies individually adapted for each one:

- Fertilizer capability: Soil levels of potassium (K), nitrogen (N) and phosphorus (P) are going to be analyzed in both samples. Results are going to be evaluated monetarily with the difference in terms of euros spent in fertilizer to reach ideal levels for crops growth by a non-user of reclaimed water farmer compared with a reclaimed water user since a great part of these costs are avoided by reclaimed water users because of the fertilizer capability of this water resource.

- Crop productivity: Kilograms of production per hectare are going to be measured in the same period of the year during 8 weeks within open season. In both farms is grown the same crop variety under same irrigation conditions in terms of quantity of water used to irrigate. Then, medium prices at which the product grown have been sold during crop season are going to be used to know in economic terms how differences in crop productivity influences farmers' profits.

- Soil quality: Two analyses in each farm are going to be carried out to evaluate the soil quality (macro and micronutrients and soil microbial communities) after being exposed during a long period of time to different water sources. The economic value will be measured using the results of the reclaimed water farm analyses and calculating how much would it cost to buy everything needed to reach the same levels for the other plot.

- Plant growth: Volumes and mass from plants grown from the same variety of seeds are going to be measured. Formation of new structures such as cells, organs or tissues are also going to be considered. Last but not least, plant development (directly related to cell and tissue specialization) and reproduction (understood as the production of new individuals) are also going to be taken into account. The economic value will be measured using the results of the reclaimed water farm analyses and calculating how much would it cost to reach same plant growth levels for the other plot.

- Source of organic matter: It is a viable and valuable source of organic carbon for soils because it provides energy to join soil particles into aggregates, promoting plant growth and enhancing soil stability. Furthermore, the organic matter also increases the good aeration of the soil and water infiltration which may not only reduce, but also even prevent a flood in the farm in case a heavy rainfall happens. Samples from both farm plots are going to be analyzed and differences between them in terms of organic matter are going to be economically measured using the market price of organic matter.

- Contribution to protect water bodies and capacity of enhancing and boosting marine ecosystem services: These two positive aspects derived from the use of reclaimed water in agriculture are inextricably linked. Treating wastewater and channeling it into new water resources avoid sea dumping thus protecting water bodies and marine ecosystem services by reducing the amount of pollutants spilled into the sea that pollutes the underwater flora and fauna. Furthermore, as the use of reclaimed water reduces the use of fertilizers, water pollution caused by fertilizers would also be reduced. Methodologies used will be:

o   Chromophoric dissolved organic matter (CDOM) monitoring: As while it is being discomposing it releases tannin which reduces the pH of the water and depletes the oxygen levels.

o   Conductivity, salinity and total dissolved solids (TDS) monitoring to see if levels may jeopardize marine ecosystem services.

o   Chlorophyll fluorescence analysis: As phosphorus and nitrogen (both fertilizers used in agriculture) may lead into an uncontrolled growth of algae which results in oxygen depletion at levels that may be toxic to the fauna and flora.

o   Biochemical oxygen demand test to see how much sewage is found in the water by counting decomposer bacteria in the water.

Using a sample of water from an area where wastewater is dumped, the economic evaluation will be done calculating the cost of cleaning and purifying the water analyzed artificially until reaching same levels as in the non-polluted water sample.

•   Promote water resources quality and recovery: When wastewater is properly treated and used for the irrigation of agricultural crops it enhances water bodies recovery since using reclaimed water means that pressure put under already overexploited wells or aquifers will be reduced so there will be a higher chance of its regeneration. Furthermore, reclaimed water will help in the improvement of the quality of water resources since reducing the level of water extraction would reduce salinity and conductivity. As if local wells or aquifers had enough quantity of quality water transfers from other water alternatives would not be necessary, using reclaimed water would avoid the cost of distribution channels from other locations and the energy cost so this is how the economic value of the reclaimed water in promoting water resources quality and recovery will be calculated.

•   Promoting agriculture: The area of study, where agriculture has an important value in terms of the contribution to the total revenues will be compared with another geographical location with very similar area (km^2), soil structure and climatic conditions (temperature, humidity, and rainfall) but with a high scarcity of water resources which impede the implementation of the agricultural sector. Then, the economic value generated in the area by the agriculture sector using reclaimed water to irrigate crops would be compared with the scenario without tertiary water.

•   Escherichia coli control: Since wastewater is a possible source of some bacteria as Escherichia coli, different concentrations of this bacteria are going to be measured in samples of treated wastewater with the minimum conditions required by European legislation for sea dumping and treated wastewater after minimum conditions needed for its use to irrigate agricultural crops. Differences in the cost of the two treatments are going to be used to see how the farmer ends up bearing the total cost of a process that if it was not done for irrigation of his crops should be covered by the government to comply with the regulations.

Furthermore, for this research reclaimed water samples will be compared with different of surface water samples from rivers which are surrounded by industries, and which are used to irrigate adjacent crops.

4 HIGHLY PREDICTABLE RESULTS

Although this study is getting going and the time to fully carry out the research will take years in order to guarantee different sampling periods under different weather conditions to avoid a bias in the results, using already existing research about some of the parameters here studied some results may be predicted with some certainty. Reclaimed water capacity of being a source of organic matter (OM) is expected to almost triple the initial samples parameters after being irrigated with reclaimed water while values of soil samples after being irrigated with well water are predicted to remain stable without relevant variations or even certain decreases in the OM concentration [40]–[42]. Conversely, another of the many benefits derived from the implementation of tertiary water to irrigate crops directly that is expected to be shown in this study is a great increase in the growth of the crops, which affects the growing process potentiating the growth in the yields of some crop varieties up to 61% [43]. Productivity also goes through relevant variations since the implementation of reclaimed water to irrigate crops may double crop production in terms of kilograms produced by the same crop variety during the same period and under same external conditions [44], [45]. Last but not least, important reductions in the use of fertilizers are expected since previous research in the area of study proved a decrease in amount of nitrogen (N) needed up to 60%, in the amount of phosphorus (P) up to 80% and in the amount potassium (K) up to 25% using as a reference the total quantities of fertilizers needed to provide crops the optimum amount to peak performance [46]–[49]. Despite that the parameters mentioned before have already been studied in other previous research, our thorough methodology will refine existing data within the same area of study so all the data can be used to study interconnections and direct relationships. Moreover, after this research different economic values will be assigned for each positive aspect derived from the use of reclaimed water in agriculture to irrigate crops which since it has not been done before it will strengthen with real and comparable data the value of such a sustainable alternative as reclaimed water is. Furthermore, there are other aspects that are going to be studied in this research for the first time ever since they have not been considered before. This make our research groundbreaking as some parameters as reclaimed water capacity of boosting marine ecosystem or its contribution on controlling e.coli concentration and distribution might show for the first time and with certain confidence the relevancy of the economic benefits derived from all the benefits of using reclaimed water to irrigate crops. Furthermore, as this research will show economic and therefore quantifiable results, all the stakeholders will take part in the boosting of the implementation of this alternative water resource in agriculture. At last, results from surface water samples might show how while this water can be used to irrigate crops without any minimum requirements or parameters and its use is widely accepted by farmers and consumers, levels of pollution are potentially way higher than in treated wastewater while farmers and consumers tend to undervalue it rejecting its use in agriculture because of the yuck factor and wrong beliefs. Feedback from highly experienced researchers after the Air and Water Pollution 2022 Conference is also a great opportunity that would provide significant added value to this research since all the suggested improvements given will be considered.

REFERENCES

[1] UNESCO World Water Assessment Programme (WWAP), The United Nations World Water Development Report 2019: Leaving no one behind. UNESCO: Paris, 2019.

[2] Deviller, G., Lundy, L. & Fatta-Kassinos, D., Recommendations to derive quality standards for chemical pollutants in reclaimed water intended for reuse in agricultural irrigation. *Chemosphere*, **240**, 2020. DOI: 10.1016/j.chemosphere.2019.124911.

[3]   Becerra-Castro, C., Lopes, A.R., Vaz-Moreira, I., Silva, E.F., Manaia, C.M. & Nunes, O.C., Wastewater reuse in irrigation: A microbiological perspective on implications in soil fertility and human and environmental health. *Environment International*, **75**, pp. 117–135, 2015. DOI: 10.1016/j.envint.2014.11.001.

[4]   Martínez-Alvarez, V., Maestre-Valero, J.F., González-Ortega, M.J., Gallego-Elvira, B. & Martin-Gorriz, B., Characterization of the agricultural supply of desalinated seawater in Southeastern Spain. *Water (Switzerland)*, **11**, 2019. DOI: 10.3390/w11061233.

[5]   Contreras, J.I., Roldán-Cañas, J., Moreno-Pérez, M.F., Gavilán, P., Lozano, D. & Baeza, R., Distribution uniformity in intensive horticultural systems of Almería and influence of the production system and water quality. *Water (Switzerland)*, **13**, pp. 1–15, 2021. DOI: 10.3390/w13020233.

[6]   Wang, X., Müller, C., Elliot, J., Mueller, N.D., Ciais, P., Jägermeyr, J., Gerber, J., Dumas, P., Wang, C., Yang, H., Li, L., Deryng, D., Folberth, C., Liu, W., Makowski, D., Olin, S., Pugh, T.A.M., Reddy, A., Schmid, E., Jeong, S., Zhou, F. & Piao, S., Global irrigation contribution to wheat and maize yield. *Nature Communications*, **12**, pp. 1–8, 2021. DOI: 10.1038/s41467-021-21498-5.

[7]   Kourgialas, N.N., A critical review of water resources in Greece: The key role of agricultural adaptation to climate-water effects. *Science of the Total Environment*, **775**, 145857, 2021. DOI: 10.1016/j.scitotenv.2021.145857.

[8]   Melián-Navarro, A. & Ruiz-Canales, A., Evaluation in carbon dioxide equivalent and chg emissions for water and energy management in water users associations: A case study in the southeast of Spain. *Water (Switzerland)*, **12**, 2020. DOI: 10.3390/w12123536.

[9]   Ashu, A.B. & Lee, S.I., The effects of climate change on the reuse of agricultural drainage water in irrigation. *KSCE Journal of Civil Engineering*, **25**, pp. 1116–1129, 2021. DOI: 10.1007/s12205-021-0004-2.

[10]   AL-Agele, H.A., Nackley, L. & Higgins, C.W., A pathway for sustainable agriculture. *Sustainability*, **13**, p. 4328, 2021. DOI: 10.3390/su13084328.

[11]   Arellano-Gonzalez, J., Aghakouchak, A., Levy, M.C., Qin, Y., Burney, J., Davis, S.J. & Moore, F.C., The adaptive benefits of agricultural water markets in California. *Environmental Research Letter*, **16**, 2021. DOI: 10.1088/1748-9326/abde5b.

[12]   Xin, Y. & Tao, F., Have the agricultural production systems in the North China Plain changed towards to climate smart agriculture since 2000? *Journal of Cleaner Production*, **299**, 126940, 2021. DOI: 10.1016/j.jclepro.2021.126940.

[13]   Pretty, J. et al., The top 100 questions of importance to the future of global agriculture. *International Journal of Agricultural Sustainability*, 8, pp. 219–236, 2010. DOI: 10.3763/ijas.2010.0534.

[14]   Hoang, L.T., Roshetko, J.M., Huu, T.P., Pagella, T. & Mai, P.N., Agroforestry: The most resilient farming system for the hilly northwest of Vietnam. *International Journal of Agriculture System*, **5**, p. 1, 2017. DOI: 10.20956/ijas.v5i1.1166.

[15]   Cabello Villarejo, V. & Madrid Lopez, C., Water use in arid rural systems and the integration of water and agricultural policies in Europe: The case of Andarax river basin. *Environment, Development and Sustainability*, **16**, pp. 957–975, 2014. DOI: 10.1007/s10668-014-9535-8.

[16]   Runhaar, H., Four critical conditions for agroecological transitions in Europe. *International Journal of Agricultural Sustainability*, **19**, pp. 227–233, 2021. DOI: 10.1080/14735903.2021.1906055.

[17] Fields, C.M., Labadie, J.W., Rohmat, F.I.W. & Johnson, L.E., Geospatial decision
      support system for ameliorating adverse impacts of irrigated agriculture on aquatic
      ecosystems. *Agricultural Water Management*, **252**, 106877, 2021.
      DOI: 10.1016/j.agwat.2021.106877.
[18] Hristov, J., Barreiro-Hurle, J., Salputra, G., Blanco, M. & Witzke, P., Reuse of treated
      water in European agriculture: Potential to address water scarcity under climate
      change. *Agricultural Water Management*, **251**, 106872, 2021.
      DOI: 10.1016/j.agwat.2021.106872.
[19] Tociu, C., Ciobotaru, I.E., Maria, C., Déak, G., Ivanov, A.A., Marcu, E., Marinescu,
      F., Savin, I. & Noor, N.M., Exhaustive approach to livestock wastewater treatment in
      irrigation purposes for a better acceptability by the public. *AIP Conference
      Proceedings*, p. 2129, 2019. DOI: 10.1063/1.5118074.
[20] Dhiman, J., Prasher, S.O., ElSayed, E., Patel, R.M., Nzediegwu, C. & Mawof, A.,
      Heavy metal uptake by wastewater irrigated potato plants grown on contaminated soil
      treated with hydrogel based amendments. *Environmental Technology and Innovation*,
      19, 100952, 2020. DOI: 10.1016/j.eti.2020.100952.
[21] Morales, M.E. & Belmonte-Urena, L.J., Theoretical research on circular economy and
      sustainability trade-offs and synergies: A bibliometric analysis. *2021 IEEE
      International Conference on Technology and Entrepreneurship, ICTE 2021*, 2021.
      DOI: 10.1109/ICTE51655.2021.9584537.
[22] Preisner, M. et al., Indicators for resource recovery monitoring within the circular
      economy model implementation in the wastewater sector. *Journal of Environmental
      Management*, **304**, 2022. DOI: 10.1016/j.jenvman.2021.114261.
[23] Chhipi-Shrestha, G., Hewage, K. & Sadiq, R., Fit-for-purpose wastewater treatment:
      Testing to implementation of decision support tool (II). *Science of the Total
      Environment*, **607–608**, pp. 403–412, 2017. DOI: 10.1016/j.scitotenv.2017.06.268.
[24] Michetti, M., Raggi, M., Guerra, E. & Viaggi, D., Interpreting farmers' perceptions of
      risks and benefits concerning waste water reuse for irrigation: A case study in Emilia-
      Romagna (Italy). *Water (Switzerland)*, **11**, 2019. DOI: 10.3390/w11010108.
[25] Baawain, M.S., Al-Mamun, A., Omidvarborna, H., Al-Sabti, A. & Choudri, B.S.,
      Public perceptions of reusing treated wastewater for urban and industrial applications:
      Challenges and opportunities. *Environment, Development and Sustainability*, **22**, pp.
      1859–1871, 2022. DOI: 10.1007/s10668-018-0266-0.
[26] Toro Sánchez, F., El uso del agua en Nijar: implicaciones ambientales del modelo
      actual de gestión. *Revista de Estudios Regionales,* **7585**(83), pp. 145–176, 2008.
[27] Aznar-Sánchez, J.A., Belmonte-Ureña, L.J. & Valera, D.L., Perceptions and
      acceptance of desalinated seawater for irrigation: A case study in the Níjar district
      (southeast Spain). *Water (Switzerland)*, **9**, 2017. DOI: 10.3390/w9060408.
[28] Jiménez-Buendía, M., Soto-Valles, F., Blaya-Ros, P.J., Toledo-Moreo, A., Domingo-
      Miguel, R. & Torres-Sánchez, R., High-density wi-fi based sensor network for
      efficient irrigation management in precision agriculture. *Applied Sciences
      (Switzerland)*, **11**, pp. 1–20, 2021. DOI: 10.3390/app11041628.
[29] Egea, F.J., Torrente, R.G. & Aguilar, A., An efficient agro-industrial complex in
      Almería (Spain): Towards an integrated and sustainable bioeconomy model. *New
      Biotechnology*, **40**, pp. 103–112, 2018. DOI: 10.1016/j.nbt.2017.06.009.
[30] Aznar-Sánchez, J.A., Belmonte-Ureña, L.J., Velasco-Muñoz, J.F. & Valera, D.L.,
      Farmers' profiles and behaviours toward desalinated seawater for irrigation: Insights
      from south-east Spain. *Journal of Cleaner Production*, **296**, 2021.
      DOI: 10.1016/j.jclepro.2021.126568.

[31]   Aznar-Sánchez, J.A., Belmonte-Ureña, L.J., Velasco-Muñoz, J.F. & Valera, D.L., Aquifer sustainability and the use of desalinated seawater for greenhouse irrigation in the Campo de Níjar, southeast Spain. *International Journal of Environmental Research and Public Health*, **16**, 2019. DOI: 10.3390/ijerph16050898.

[32]   Lo-Iacono-Ferreira, V.G., Viñoles-Cebolla, R., Bastante-Ceca, M.J. & Capuz-Rizo, S.F., Transport of Spanish fruit and vegetables in cardboard boxes: A carbon footprint analysis. *Journal of Cleaner Production*, **244**, 2020. DOI: 10.1016/j.jclepro.2019.118784.

[33]   Rodríguez Martín, J.A., Ramos-Miras, J.J., Boluda, R. & Gil, C., Spatial relations of heavy metals in arable and greenhouse soils of a Mediterranean environment region (Spain). *Geoderma*, **200–201**, pp. 180–188, 2013. DOI: 10.1016/j.geoderma.2013.02.014.

[34]   Reca, J., Trillo, C., Sánchez, J.A., Martínez, J. & Valera, D., Optimization model for on-farm irrigation management of Mediterranean greenhouse crops using desalinated and saline water from different sources. *Agricultural Systems*, 166, pp. 173–183, 2018. DOI: 10.1016/j.agsy.2018.02.004.

[35]   Honoré, M.N., Belmonte-Ureña, L.J., Navarro-Velasco, A. & Camacho-Ferre, F., Profit analysis of papaya crops under greenhouses as an alternative to traditional intensive horticulture in Southeast Spain. *International Journal of Environmental Research and Public Health*, **16**, 2019. DOI: 10.3390/ijerph16162908.

[36]   Thompson, R.B., Padilla, F.M., Peña-Fleitas, M.T. & Gallardo, M., Reducing nitrate leaching losses from vegetable production in Mediterranean greenhouses. *Acta Horticulturae*, **1268**, pp. 105–117, 2020. DOI: 10.17660/ActaHortic.2020.1268.14.

[37]   Cajamar, Análisis de la Campaña Hortofrutícola de Almería, Campaña 2019/2020, 2020.   https://www.plataformatierra.es/detalle/analisis-campana-hortofruticola. Accessed on: 19 Jun. 2021.

[38]   Consejería de Agricultura, Ganadería, Pesca y Desarrollo Sostenible de la Junta de Andalucía. Secretaría General de Agricultura y Alimentación. Cartografía de Invernaderos en Almería, Granada y Málaga, 2019. https://www.juntadeandalucia.es/export/drupaljda/producto_estadistica/19/06/Cartografia%20_inv_AL_GR_MA_190 926.pdf. Accessed on: 19 Apr. 2021.

[39]   Garcia-Caparros, P., Contreras, J.I., Baeza, R., Segura, M.L. & Lao, M.T., Integral management of irrigation water in intensive horticultural systems of Almería. *Sustainability (Switzerland)*, 9, pp. 1–21, 2017. DOI: 10.3390/su9122271.

[40]   Mohammad Rusan, M.J., Hinnawi, S. & Rousan, L., Long term effect of wastewater irrigation of forage crops on soil and plant quality parameters. *Desalination*, **215**, pp. 143–152, 2007. DOI:10.1016/j.desal.2006.10.032.

[41]   Yasmeen, T., Ali, Q., Islam, F., Noman, A., Akram, M.S. & Javed, M.T., Biologically treated wastewater fertigation induced growth and yield enhancement effects in Vigna radiata L. *Agricultural Water Management*, **146**, pp. 124–130, 2014. DOI: 10.1016/j.agwat.2014.07.025.

[42]   Bernier, M.H., Levy, G.J., Fine, P. & Borisover, M., Organic matter composition in soils irrigated with treated wastewater: FT-IR spectroscopic analysis of bulk soil samples. *Geoderma*, **209–210**, pp. 233–240, 2013. DOI: 10.1016/j.geoderma.2013.06.017.

[43]   Wang, Z., Li, J. & Li, Y., Using reclaimed water for agricultural and landscape irrigation in China: A review. *Irrigation and Drainage*, **66**, pp. 672–686, 2017. DOI: 10.1002/ird.2129.

[44] Paul, M., Negahban-Azar, M. & Shirmohammadi, A., Assessing crop water productivity under different irrigation scenarios in the mid-Atlantic region. *Water (Switzerland)*, **13**, 2021. DOI: 10.3390/w13131826.

[45] Palese, A.M., Pasquale, V., Celano, G., Figliuolo, G., Masi, S. & Xiloyannis, C., Irrigation of olive groves in southern Italy with treated municipal wastewater: Effects on microbiological quality of soil and fruits. *Agriculture, Ecosystems and Environment*, **129**, pp. 43–51, 2009. DOI: 10.1016/j.agee.2008.07.003.

[46] Athamenh, B.M., Salem, N.M., El-Zuraiqi, S.M., Suleiman, W. & Rusan, M.J., Combined land application of treated wastewater and biosolids enhances crop production and soil fertility. *Desalination and Water Treatment*, **53**, pp. 3283–3294, 2015. DOI: 10.1080/19443994.2014.933037.

[47] Mohammad Rusan, M.J., Hinnawi, S. & Rousan, L., Long term effect of wastewater irrigation of forage crops on soil and plant quality parameters. *Desalination*, **215**, pp. 143–152, 2007. DOI: 10.1016/j.desal.2006.10.032.

[48] López-Serrano, M.J., Velasco-Muñoz, J.F., Aznar-Sánchez, J.A. & Román-Sánchez, I.M., Economic analysis of the use of reclaimed water in agriculture in southeastern Spain, a Mediterranean region. *Agronomy*, **11**, 2021.
DOI: 10.3390/agronomy11112218.

[49] Segura Pérez, M.L., IFAPA, Centro La Mojonera. Jornada de presentación de las actividades de Investigación y Trasnferencia relacionadas con la Reutilización de Aguas Regeneradas en Cultivo de Pimiento. Proyecto INIA. RTA2006-00032-00-00, Almería, 2010.

Author index

Adami L. 53
Akter A. 99
Arauz E. 87

Baeza-Serrano Á. 137
Briones-Bitar J. 125

Carrión-Mero P. 125
Chamon A. S. 99
Chau P. N. 25
Cheung H. Y. W. 75

de Luque-Villa M. A. 65, 111
Dominici A. 87

Faiz S. M. A. 99
Fayos G. 137
Fernández-Garza A. G. 37
Fraiz A. 87

Gielen E. 37
González-Méndez M. 111
Gonzalez-Rugel A. 125
Granda-Rodriguez H. D. 65
Gutiérrez P. 137

Ho T. 15

López Serrano M. J. 147
Lu J. 15

Malki-Epshtein L. 75
Merchán-Sanmartin B. 125
Mondol M. N. 99
Morante-Carballo F. 125

Oliver N. 137
Ortega-Samaniego Q. M. 87
Osorio H. 87

Paches M. 87
Palencia-Jiménez J.-S. 37
Parra R. 3

Rada E. C. 53
Ramos-Merchante A. 87
Rodriguez N. 15
Romero I. 87
Romero-Tovar G. 65
Rybarczyk Y. 25

Schiavon M. 53
Sempere F. 137
Shi Z. 15

Tubino M. 53

Vera-Demera H. 125

Zalakeviciute R. 25

WIT*PRESS* ...for scientists by scientists

Urban Water Systems & Floods IV

Edited by: **S. MAMBRETTI**, *Polytechnic of Milan, Italy and* **D. PROVERBS**, *University of Wolverhampton, UK*

Research works were presented at the 8th International Conference on Flood and Urban Water Management with the aim of developing innovative solutions that can help bring about multiple benefits toward achieving integrated flood risk and urban water management strategies and policy. The papers resulting from these works form this book.

Flooding is a global phenomenon that claims numerous lives worldwide each year. When flooding occurs in urban areas, it can cause substantial damage to property as well as threatening human life. In addition, many more people must endure the homelessness, upset and disruption that are left in the wake of floods. The increased frequency of flooding in the last few years, coupled with climate change predictions and urban development, suggest that these impacts are set to worsen in the future. How we respond and importantly, adapt to these challenges is key to developing our long term resilience at the property, community and city scale.

As our cities continue to expand, their urban infrastructures need to be re-evaluated and adapted to new requirements related to the increase in population and the growing areas under urbanization. We also need to consider more nature-based interventions to the management of flood risk, including the adoption of more catchment-based approaches. These are now being recognised as being more sustainable and also able to achieve wider benefits to the environment and society as a whole.

Water supply systems and urban drainage are also increasingly important due to this expansion. Topics such as contamination and pollution discharges in urban water bodies, as well as the monitoring of water recycling systems are currently receiving a great deal of attention from researchers and professional engineers working in the water industry. Mitigating losses from water distribution networks and effective, efficient and energy-saving management are key goals for optimising performance and reducing negative impacts. Sewer systems are under constant pressure due to growing urbanization and climate change, and the environmental impact caused by urban drainage overflows is related to both water quantity and water quality.

This book is aimed at researchers, academics and practitioners involved in research and development activities across a wide range of technical and management topics related to urban water and flooding and its impacts on communities, property and people.

WIT Transactions on the Built Environment, vol. 208

ISBN: 978-1-78466-469-5   eISBN: 978-1-78466-470-1
ISSN (print): 1746-4498   ISSN (online): 1743-3509

Published 2022 / 144pp

www.ingramcontent.com/pod-product-compliance
Lightning Source LLC
Chambersburg PA
CBHW062008190326
41458CB00009B/3004

* 9 7 8 1 7 8 4 6 6 4 6 7 1 *